The Two Himalayas

Fred Apps

The Lutterworth Press
Cambridge

The Lutterworth Press
P.O. Box 60
Cambridge
CB1 2NT

British Library Cataloguing-in-Publication Data
A catalogue record for this book is available from the British Library

ISBN 0-7188-2866-6

Copyright © Fred Apps 1992

First published by The Lutterworth Press 1992

All rights reserved. No part of this publication may be reproduced,
stored in a retrieval system, or transmitted in any form or by any means,
electronic, mechanical, photocopying, recording or otherwise,
without the prior permission in writing of the publisher.

Printed in Great Britain by
WBC Print Ltd., Bridgend, Mid Glamorgan

1

They burnt a man the night I stayed up late in Mr Guy's field. That's what I thought at the time, but it didn't worry me. It was 11 November 1918, Armistice Day, and I was not quite five years old. I was just a bit puzzled as to whether he liked it or not.

I didn't know that the "burning man" was a pretend Kaiser, something to do with a war, the worst ever war.

I thought big fires were caused by Zeppelins - like the one at Southend. But it went out special so that I could be taken there to collect winkles in the mud. I remember the horse-cab coming special to take us to a station: and it having to turn round special so that I could collect my forgotten Teddy Bear.

But I'd since grown up bigger and got a sailor suit, with 'HMS *Lion*' round the hat, for Ilford Station - where I sometimes met Daddy - and where the dark, wet, windy station was the bridge of some big Dreadnought . . .

Except that the time we went to Liverpool Street, where all the trains were, he lifted me up shivering from the big hissing steam engines, because Sailors don't cry.

Ladies always said I mustn't cry, but they fixed their hats to their heads by pushing a long shiny pin in one side and out the other so that it went right through their heads. Nor could I make out about the Germans, who'd taken Uncle's foot off - I think in the mud . . . but why were they under Southend pier? Nor about the murder next door. The Auntie there had got a boy-friend (whatever that meant. Worse than murdering!). I thought the hole in the fence - my hole - had got mended special, 'cause he got through it too.

I didn't understand boy-friends - but they seemed to make murders even more special.

I remember the old trams up to Aldgate. In the Transport Museum now they seem the wrong size - the driver out in the exposed cab looked high above me then. I didn't think about how cold he must have been, open to the weather all day, with only a waist-high screen; stamping desperately as he moved that magic handle that made the tram whine faster. What grinding pressures made him stick *that* job?

I remember the poor horses lying tangled in their shafts, unable to regain their feet, having struggled with a brewer's dray along the slippery, cambered, cobbled Whitechapel roads. All so exhaustingly unmechanized. The only place for petrol between London and Brighton was a grocer's shop in Redhill. Old Mr Coleman used to stop there with his wondered-at motor bike, which almost qualified him to be a pilot over the First World-War trenches.

All so much more than a thousand miles away from Himalayan India . . . though memories get tangled.

The Two Himalayas

Somewhere between the mangel-wurzles of those war years and the plum-puddings of my youth, was my first piece of butter, which I remember clearly, one Sunday afternoon. Because dad had got to be senior clerk.

He'd started work in an East-end stables, where the treatment was pretty rough. So he'd made up his mind to get fit, "so as to be able to look after himself." He must have worked hard to get away from all that... and played hard too, because one year he won the London to Brighton roadwalk.

But in 1914 the Army turned him down because of a 'weak heart'. So he shovelled coal three or four hours a night, unpaid voluntary, into the Power Station boilers... and eventually brought that piece of butter home. Sunday afternoon, perhaps with a bit of jam.

I wasn't really old enough to know enough about those years - years when just across the water, the relentless leaving-nothing sacred of the flies.... It had been 'good for the men's morale' to carry seventy pounds of belongings over the top so that they could hardly stagger through the uphill mud, and slid and fell on the downhill mud. Wounds, barbed-wire, flies... and the night air out there tortured with a sound 'like an enormous wet finger screeching across an enormous pane of glass'.

That was the Somme, and there lay their belongings: a man's morale clutching at the mud... with quieter sounds.

I didn't realize that *my* dad might easily not have been at home. I can only remember our garden: that old wooden box which was a ship, then an airship, then a bus. A few nails for masts - or for guns....

But years later, the Two Minutes' Silence tore at my guts; at all our guts.

I belonged to St Margaret's Church Choir. We got a penny a week which we were reluctantly persuaded to spend on laundry for our white collars: a well-meant parental iniquity that really tangled up our loyalties. If the choir provided godliness, someone else should fix the cleanliness, like they did the clean knees. We guessed nothing of the Somme - and how they'd had to try and heal themselves with the little things of ordinary life; and perhaps hidden, kneeling things.

Though our real interest was racing woolly caterpillars across to the opposite row of choirboys, I believe that some awareness of what the Church was supposed to stand for rubbed off on me. A love for the beautiful charismatic language of the parables, and some stirrings from the meaning of the Bible stories still remains, rooted in those days.

Sadly, though, I grew into a worried tangle over the conviction that I was sinning nightly against the Holy Ghost. The chanted words 'Take not thy Holy Spirit from me' - deliberate and muted - haunted me uncannily, alone with my guilt, sinking, unforgiven.

Something terrible was happening to me; especially to me. I prayed nightly - an intense penance that somehow slid away into an abyss: those Holy words crying for unholy me. But forgiveness would come, and release me back to strangely untroubled fantasies about some mysterious feminine world, and lovely sleep-bringing masturbation.

Until one day, one night I expect, I suddenly saw it as the everlasting futility that it was. "Bugger the whole thing!" I thought, in the strongest language I knew. And I distinctly remember the tensions exploding; I was off the hook that I should never have been on anyway. I consigned it all to the devil and got to falling off to sleep with or without.

The world could hardly have got itself in a worse mess - children a mystery to their parents, and the parents a mystery to their children. Not without considerable help from the Church of those days, I expect.

Nowadays, when it's all more accepted as natural, I imagine a father giving an ever-so-gentle wink; and a little understanding smile light up a bond between them - yet still maybe not be as worthwhile a bloke as my dad was.

There was another problem in my schooling holding me back from friendships, with their latent perspectives. An improbable, almost unrecognizable complaint afflicted me - an inability to catch a ball. It could not have been known by me, nor the boys in the suburbs at large, that eye-muscles need sufficient blood-sugar to focus a quickly-moving object: which I didn't have. I only had eyes to sense the shame, the derision - and a finger pointing at my 'nerves'.

Not only were cricket and tennis out, but I also lived in constant fear of getting involved with groups of boys whose constant contempt and derision - exaggerated in my mind of course - were quite unbearable.

It sounds strange, but I would have given anything to have been able to claim some sort of obvious disability that could have let me honourably keep their company without being expected to catch a ball. Suburban life was tied and limited to the conventional. I even longed at times to be a cripple, secretly envying a boy in my class who only had one leg and from whom no sports were demanded.

I became shy and introspective. I cycled miles by myself, longing for some adventure. I had no friends, though the wilder, more unruly spirits in my class at school would talk to me. I suppose they sensed that there was an echo of their natures hidden somewhere in mine. I tried to live up to their friendlier attitude by doing silly things, bating authority in stupid ways - but ways that wouldn't risk involvement in a fight, which I would have lost, and so lost their regard by making a fool of myself.

I must have been far unhappier than I care to think about. I took to dreaming about the sea. Although none of my relations had, as far as I can make out, ever done more than paddle for an hour or two, I think I must have the sea in my blood. Or perhaps the yearning had been stirred by my childhood reading, and by vivid childhood imaginings.

My uncle down the road added fuel to the fire. He had lots of big pictorial volumes of naval history. A kind-hearted suburban man, he played host to us kids at Christmas. My younger brother and I used to get great amusement from watching his bald head perspire, because of the horse-radish sauce, when he carved the joint. But I suppose we grew out of this and struggled to read those books.

How old are you when you only ever read things kneeling on the floor? Ten? Eleven? It must have been a highly impressionable and imaginative age because, give and take a bit of boyish misinterpretation, I still have mental pictures of those vividly daring breakaways from the accepted line of battle that made the 'Nelson touch' such an inspiration: stirring even to the poor, half-starved, press-ganged, crews and the illiterate pub crowds of the time, they struck a chord in my spirit and heaped coals on my frustrations.

Nelson's daring was somehow in the magic of a ship's tall rigging, and in the first rope that old Sam, the Leigh-on-Sea fisherman, let me hold. This idol of my hero-worship came out of a real sea-dog mould, even though the most adventurous thing I'd ever seen him do was to give the holiday-makers a quick ride in his little sailing boat. The rope he handed me - I remember it still - would probably have been flopping idly on the deck where it belonged if I had not held it; but I thrilled and nearly burst with pride and salty, pregnant wisdom.

Although many thousands of people have got into more sailing adventures and experiences than I ever have, a boat was the only place where, from the very beginning, I felt some kind of inborn competence. To get out of every scrape with ears only to my own intuition, with the same confidence with which I had got into it: ever seeking the challenge, and not the advice of those with engines! Excusable, perhaps, for younger days, but of course it's 'all wrong', not really, 'Yacht-Club'. Yet keeping out of trouble by switching on the engine is one good way to make sure you never learn to get out of trouble on the day you have to rely on sail alone. Your Yacht Club blazer won't help you then.

I was never really strong enough physically - nor did I have the money - to try those standard tough deep-sea exploits. Sandbanks, tides and tricky coasts really attracted me more anyway. But I dreamed, and not long ago still could do, of the exploits of lifeboatmen. Something was vaguely wrong with my life, that I couldn't be one.

The old Thames spritsail barges out in the distant fairway fascinated me with their dark red sails. The sprit is a diagonal spar that avoids an unmanageably heavy boom at the foot of the mainsail, so that a man and a boy commonly handled these lovely craft. I wanted to be one of those boys.

I was promised a coastal trip on one once. I cycled miles, home and then back again for the next tide, but he'd left - thinking perhaps that I was too young. I know I wasn't old enough for those hurting, disillusioned miles cycling back home.

Later - much later - I saw several of these old barges moored abreast alongside the quay at Maldon in Essex. They lay in moonlight, and in a flash I was back in the days when I longed to learn the secrets of their uncanny navigation.

Having no friends, I had to amuse myself. I buried my loneliness in an ambition to build a model barge. It was quite big, made up of separate little planks. The hull was designed from a half-understood library book, the ribs all shaped from plotted curves that came from a fourteen-year-old's struggle with the grown-up world of the drawing board. The boat still stands in my bedroom, with its deep, fine, trawler-shaped hull and exact Thames barge rigging: a nautical incongruity (barges being square for cargo). But I loved the deep trawlers, and I loved the square sailing barges, and they both inhabited the same dream.

Just beside it stands the '*Fighting Téméraire*', Turner's famous picture of the last voyage of that lovely old square-rigger, with the black tug fussing her out of her past, and the sadness in the background sunset. That's how they will stay, next to each other, while they are in my room.

I had left school in a half-studious, half-bored frame of mind, with no particular regrets as far as I can remember. Jobs were hard to get at the time; I nearly landed up as a lamp-

The Two Himalayas

a lamp-black manufacturer's assistant. I think my mother must have been secretly relieved that somebody else got that job, because washing was hard work in those days. Even vacuum cleaners were a bit unusual. My dad apparently experimented with such things, new-fangled in such surrounding ignorance. The lady opposite went over her whole house with hers without knowing that it had to be switched on. And the 'daily' emery-papered our new stainless steel cooker, never having seen that sort of bright metal in any kitchen before. Its shine never came back.

I went for an interview in the Westminster City Council Clerk's Department (I suppose that my dad's boss knew somebody there). The place - just the name 'Westminster' - scared me. I remember regarding even a teenaged junior as some sort of God. I never did have to bow to him, though, since I failed to get the job.

Eventually, through some little connection of the not yet nationalized, unamalgamated, Stepney Electricity Undertaking, I got into the offices of a cable company. My initial visions of cable-laying ships or of chatting with farmers about new overhead lines soon disappeared. Many contracts were abandoned and staff put off in the recession that hit trade just at that time. I did get into something worth doing in the end, but until then I was stuck, for three painful years, in their Inwards Invoices Department in the City.

This office life was a specially refined form of misery. The work itself was boring in the extreme. The boss was a nice, meek little man, whose legs wouldn't steer him between two desks without bumping into one of them. Even so, any confusion in the Invoice Register would terrify me out of a gloom - but one in which I knew my place: the least confident, half-accepted, lone day-dreamer amongst the invoices, a mind filled with thoughts of sailing around the world. And then one evening my lonely youth all burst. My dad bought me a little seventeen and a half foot sailing boat, *Annabelle*. It cost, I remember, exactly £1 per foot.

From then on I began to find some self-respect.

I dreamt more and more of boats and some wonderful girl companion. The pale-faced, dark-eyed despatch girl mesmerized me from the corridor, and spoke to me once. 'Red sails in the Sunset, way over the sea' sang continually through my mind; and my longings about her were haunted by the tune, and its sad horizons.

In contrast to the glimpses of this fantasy figure there was, daily sitting opposite, the opposite of my day-dreams. But for the fact that she was the director's niece, this waddly female would have been long-retired. As it was, she was my immediate boss and filled my field of vision every working day - the nearest thing there has been to a cure for my romantic vision of the opposite sex.

That office was all so very cruel to young longings. When monotony overcame the terrors of making a fool of myself, day-dreams would cloud the invoices and sometimes drift me - paradoxically enough - to the fearless deck of one of my heroes, who never questioned my courage as, sinking, we fought incredible odds, against enemies who weren't really seamen and somehow fouled each other, so that we didn't sink, but superbly manoeuvred ourselves under their clumsy sterns.

Or perhaps, like Sir Richard Grenville's 'little *Revenge*', we
> ... went down by the island crags,
> To be lost evermore in the Main.

The Two Himalayas

Then Monday morning would come, and bring its own special hell. The 'little *Revenge*' would be forgotten as I started to collect the towels and reissue clean ones on the top floor of the building. I never managed this without making a fool of myself, because I was too shy and nervous to keep count and to know what I was doing. Some of the wags were quick to spot my confusion, and played cruel and nasty little tricks.

The telephone was another area of disaster. It could be that phones were not so clear and distinct as they are nowadays, but I suspect that shyness paralysed my hearing somehow or other to the point where I just didn't know what the call was all about.

It must have been embarrassing for everybody, watching my flinching from those indecipherable words: but on the whole the few men in that little department must have been a very kind and understanding lot. Even when my complete sexual ignorance became apparent they were really very decent. I remember my confusion when it was discovered that I had never heard the word 'cunt' and obviously didn't know anything at all, except that I was painfully ignorant. They gave me one or two little leads which failed completely, and very kindly didn't shake their heads. That part of a girl's body was an abstract, magnetic confusion where babies somehow slid out, and not for me to know about.

It may seem hard to believe for today's more enlightened youth; but readers will just have to take my word for it.

My very first clue was a wisp of escaping hair at a little country swimming pool. I must have been getting on for twenty years old. The wisp transfixed me, and resulted in two or three visits to the local libraries to hunt for what it could mean.

I didn't get very far. One misty quarter frontal would have had any book banned in those days.

As a result of this desolate, insidious perplexity I had in later years no hesitation in taking my two children to a naturist beach when they were quite young; and they spent most of their free time and school holidays swimming in a naturist pool, near where we lived, getting fit and playing out of their system any future problems and the sort of miserable turmoil I had endured.

But four nights a week at evening classes - desperate to get another job - didn't help with *my* bottled-up problems. So that I dreamt night after night of accidentally-embarrassed topless maidens, stranded in country pool or tidal estuary, exquisitely abandoned to my chivalry. It was perhaps fortunate that I could imagine *that* much of them, because of mermaids or something, not shrouded from my mind like the vague erotic mystery down *there* that was so intensely near and yet so impenetrably tabooed. Perhaps my dreams were a kind of sexual brinkmanship; growing-up restlessness and chastity all confused.

Maybe my bouts of religious idealism were the source of that erotic chivalry. Yet the religious prompting was also confused, for I had a recurring desire for that profound solace which I connected with the New Testament, so beautifully and poignantly held out, even (I supposed) to crippled social lives like mine.

I was really looking for some alternative to the normal world of friendly groups of boys: but I couldn't find anything that worked out as the Minister said it was supposed to. Or was it that I wanted to find some prop for my weak personality with which I could one day impress the girl of my dreams?

The Two Himalayas

I was now at that special age - almost the only period of my life - when a few odd lines of poetry seemed to hit a chord, and to take root inside one. Not only the chord of romantic adventure, but also some deeper elusive longing of one's teenaged years. I mean that poetry-awakened, rather haunting idea that is stirred up and means something, so that even when overtaken by our growing up it stays in our subconscious for life.

The strange thing is that many years later I was sitting in the kitchen, the family kitchen, one evening a few months before he left. He must have been about that same age, my son. We called him 'Ulli', short for Ulrich. The name is German - the reason will become clear.

He was restless, soon to wander out to India. I was vaguely aware that, rather untypically, he had got hold of an old school poetry-book of mine: "What on earth?", I thought . . . but he suddenly looked up and said, "Dad, listen to this." I suppose my eyes made him pause - I switched back many years. Out of the blue, surprised at myself, I guessed . . . .

"And I have felt
A presence that disturbs me with the joy
Of elevated thoughts; a sense sublime
Of something far more deeply interfused,
Whose dwelling is the . . ."

I petered out.

"Whose dwelling is the . . . , something, something . . . the setting sun, and in the mind of man . . . "

"Not 'the something, something', Dad," he prompted.

" . . . Whose dwelling is the light of setting suns,
And the round ocean and the living air,
And the blue sky, and in the mind of man . . ."

He paused, while I slowly remembered . . . . "That's right," he said, "but however did you hit on it just like that - didn't think you knew any poetry?"

"Three or four other bits, perhaps, I suppose."

The father in me was pleased. Those same few lines from Wordsworth had 'got' us both. And now, after all that happened later, the memory of that evening in the kitchen with its sudden, unexpected, empathy between us, brings a lump to my throat; like that I got at the end of the Live Aid concert at Wembley, thinking of his own 'live aid' longings that came to nothing. He would have felt the spirit of that movement, and probably, too, the less well-understood, restless, half-religious idealism of my own youth . . .

"What are the other bits, Dad . . . ?"

"In Xanadu did KUBLA KHAN
A stately pleasure-dome decree:
Where ALPH, the sacred river, ran
Through caverns measureless to man
Down to a sunless sea."

He gazed out of the window, just as I had, a day-dreaming invoice clerk, silently reciting it to myself in Holborn, until the interrupted pedestrian cares and thoughts of daily life closed back around me.

Not long after, when he was sick, I found him in a Temple by the same Himalayan river that had so haunted Coleridge. When Dr Singh's dysentery treatment had got him on the road to recovery, it was there in the hills that I left him.

I can still see him - finally back in London - he who had twice allowed a Holy Man to cut an abscess out of his foot without any drugs - reeling backwards from a spoonful of water: or dragging himself from the hospital bed to collapse under the sink for fear of drowning in his own saliva.

There had been no warning; no dog had bitten him. Just sudden, weird, alienating behaviour, continuous discomfort, a sleepless night. Then this mind-blowing water-terror, that most hideous sign of rabies. Increasing spasms; choking for breath. Two more sleepless nights, with just short periods free to think it all over, to wonder about sanity, about choking to death next spasm, about death itself.

Perhaps it was better for him not to know that it is still called the Rage - 'La Rage' - nor to have heard of the ancient need to confine sufferers in cages? However that may be, I have this note in front of me, shaky but clear - a last few words:

'In case you don't make it in time, and in the event of this body's death, a short note: Nothing much to say - will entrust myself to God, where I know I will continue from bliss to bliss, hoping you will . . . one day follow . . . .
I know I'm OK . . . Don't worry . . . .
    Love,
        Ulli'

You may shake your head, and dilute it for your comfort. But the message was sacred in the weakness . . . and there was that other little note with its mysterious confidence - compellingly unorthodox - 'He who grieves for me does not know Me.'

Does not know *Me*?

That had to do with his pilgrimage, coming out of a background of turbulent, restless years. So that struggling to survive amongst those last mindless, mind-searing days, he had found some untrammelled, naked sense of God - groping back to a 'sense of something far more deeply interfused' - from where Alph, that Sacred River ran? So this is a sad, yet perhaps not quite so sad, part of the story.

Not quite a conventional theme, you may agree: and not a conventional hero for a half-spiritual tale. He smoked and swore, and pilgrims don't do that. I felt my youth had been, by comparison with his crowded pilgrim story, a terrible waste. That youth of mine, sexual and spiritual hunger aching and dreaming each other to sleep, or not to sleep, and never to sail away. Wanderlust-induced inertia, contradictory and complicated and getting nowhere.

I should have read Somerset Maugham's *The Razors Edge*. It held something of the truth for me. But I waited in a way until years after the war and after my own time near the Himalayas when, back home, that story hit me from a little suburban cinema screen and touched off the nerve of this yarn for me. Because Maugham's valley was real; was in fact that same lonely Guru Shri Meher Baba's valley that I turned back from when instincts told me to go - to eat one more meal with the 'criminal tribes', and go.

And all of this has strange echoes of that sacred river of Kubla Khan and of our Temple memories, Ulli's and mine. The atmosphere of the region, the distant feel of Coleridge's poem singing clearly in the river there, at the bottom of the valley; the evolving sense of spiritual purpose; these were enough to find the same susceptibility in me and in my son, even though parts of our stories were separated by a generation of years and half a mountain range of miles.

Perhaps, because of all this, because of these coincidences and the tragedy of my son, I'm especially impressionable; but surely the ultimate and most refined intertwining of the spiritual and the sexual - of their fusing into an inner mission - is to be found in *The Razor's Edge*? In that story, Larry, abandoning a career for a life of poverty and attic reading, went to India where, alone on the razor's edge of the Himalayan range he felt compelled to go to this special particular valley to see Guru Baba. To be told that the truth of his life was to follow the Good Samaritan's example down on the Marseilles waterfront. He became, to use a modern phrase, 'downwardly mobile'. 'Lose and you will find'; he renounced his previous life to rekindle hope in a long-since prostituted girl from his past.

For me, remembering this story seems always to induce a strange feeling. There was a time when it used to come back and buzz my inner core, haunting me; homing on to some never-ever tuned-into, inner core of *me*. Just like Larry, I used to get this 'Razor's Edge' feeling: the notion of renouncing *now* so that spiritually, in some deep and unforeseeable way, there may be a final gaining.

The girl in my case was scarred not by prostitution but by sadness; and I had very little doubt that I had awoken hopes in her, and then left her deserted. And I have very little doubt that those earlier sleepless dreams of chivalry and romance were combining with my shame. So that, to atone for the harm I had done to her, this native, naïve girl from Ireland became in a way my special 'Razor's Edge', and meant so much to me.

I don't suppose Ulli ever saw that film, yet that whole story seems to have a timeless ring: vivid reactions from my old Himalayas seeming to harmonize with the mysterious truth-seeking compulsions of his life. I get something of the same feeling whenever I'm a little bit stirred by television, with its occasional wincing from Ethiopia and flies crawling in the children's eyes . . . .

2

But to get back to that youth of mine, so preoccupied with problems it couldn't solve. I did have my little boat, *Annabelle*. It was only seventeen feet long - twelve inches for every year of my life - and was a crazy vessel by today's standards. It depended upon an annual coat of tar to stop it leaking too disastrously under the strain of heeling over a bit. Even waves less than two feet high would make her so unlively and so unresponsive that you had to get an oar out to tack.

I kept it in little anchorages of the Thames area, but laid her up in winter at Twickenham, near my new digs. The swotting had paid off. I was by then a very junior draughtsman in the electricity supply offices of that district, earning just over £1 per week. I could afford tar, but not paint, on *Annabelle*.

Taking her up and down the Thames with its twenty-seven bridges was rather idiotic, something of a test of resourcefulness especially when drifting where you didn't ought to be. I found that a half-open boat trapped under a jetty where the water sluiced itself into foam still somehow floated, the foam-level a fascinatingly constant but rather mesmerising two or three inches below the gunwale.

Death was quite near a couple of times, dark and wet, or by a thrashing propeller in the sunshine, with no one to see what had happened on either occasion.

My brother Bob was with me the first time. We had rather unwisely gone ashore late one night down the river near Erith to phone an anxious mum. As we pushed off into the dark, what we had taken to be the outline of the Essex shore suddenly resolved itself into a nearby line of empty lighters - those big barges that the tugs tow around. They were moored abreast, all tied to one buoy, blocking our tide-borne drift up a river that was flooding round the bend.

The week before a Twickenham man had drowned, sucked under one of these things, his mast having hit the high overhanging prow which sticks out at forty-five degrees.

We couldn't see the barrel-shaped buoy, but by some fluke we drifted on to it so exactly that we lay sideways to the tide, see-sawing gently in time with the obscure mechanics of water rushing past, the widest part of the hull pivoting slowly on the widest part of the buoy. I remember a strange phosphoresence and the dark mental cloud of last week's accident, my amazement at this freak situation deadening my mind. Unable to think, and unable to wait any longer for thought, I suddenly grabbed an oar. Knowing, but with a half-paralysed logic, that it was impatient folly, I pushed it against the buoy... and I lay awake most of that night uneasily going over why that oar's slippery push in the phospherescent darkness got us clear of those nasty black shapes. That tide should have got us and sucked us under, filling up at forty-five degrees. Oars just don't push a boat right off clear of everything in one go; something made purchase possible. Our two lives must have pivoted unthinkably on the algae or the rust or the scratched-off paint on that curving old barrel's surface. Luck, I suppose; the sort that looks after drunks, and the sons of worried mothers.

The second time I was alone. The tide was flooding gently up the river. There was no wind, and of course I had no engine.

I had noticed that the bigger ships seemed to carry odd bits of flotsam and jetsam along with them; so I manoeuvred my little boat by rowing it in behind one or two of the bigger vessels and glided in the slipstream. This very reckless piece of nautical foolishness kept me amused and very self-satisfied for a time.

She was riding very high out of the water, the screw barely half-immersed and sending columns of spray under the high overhanging counter. Very fascinating. But this particular vessel, not really surprisingly, had reached the end of her voyage - the Royal Albert Docks. The propeller suddenly stopped, then started reversing. I shouted desperately, but no friendly figure peered over the huge towering stern. *Annabelle* was being sucked nearer. It did not occur to my bemused brain that the ship was just checking her forward motion. All I saw, and I still remember it very clearly, was the patch of water between *Annabelle* and the liner gradually diminishing as I accelerated towards the thrashing propeller.

As I watched, it suddenly stopped; but this nasty incident frightened me badly, and I learnt a searing lesson about avoiding manoeuvering ships, and ships in confined waterways generally.

When you look over London Bridge at night it is easy enough to shake your head and cleverly and wisely purse your lips at the thought of a sailing-boat with no engine, even in daylight, fouling unsuspecting liners. A stupid thing to do? Yes, but I was only seventeen years old.

I realize too, of course, that this is not in the same category as real yarns of the sea, of plunging into storms, the old tough days, and all that. But then, this is primarily about escaping from the cruel (or even crueller) life of shyness in suburbia. That motive power which drove me, short of cash as I was, to sail at night at all costs just because that is what *real* seamen do. Accommodating the rules of the sea with a paraffin lamp and two bits of cellophane, one red and one green, showing to other seafarers the colour I knew they ought to see... except that the cellophane dissolved the first time it rained. Rather sad, I think.

One final little incident of those times I would like to mention, because it might interest river-minded Londoners. It concerns the ornamental lions' heads, with their ornamental rings, which are an almost unconsciously accepted feature of the Victoria Embankment riverside wall. You can't tie up to them - and of course I had. Under some obscure bye-law, the Port of London Authority River Police asked me to untie, and towed me out amongst the moored-up barges. All very polite, except that I got taken to Court about it. Whereupon A.P. Herbert, the well-known writer, who was to become an MP and guardian of this riverside constituency, took the matter up in *Punch*, putting the view that official energies could better be spent sorting out the decaying South Bank of the river.

A strange crop of ancient riverside privileges concerning navigation, landing, drying out and so on, suddenly found themselves being exercised by self-appointed guardians of some ancient law or other. It was said that a mariner (me, I suppose) could tie up to the King's door-knocker if in distress, and that the County Council had better watch where it was meddling and leave their river alone.

What had happened in fact, was that the tide had turned and the presence of some rather bored onlookers made it impossible for me to pee, so I'd tied up to one of those old rings and sailed out to the other end of the rope. The equally bored police couldn't make it out; there was *Annabelle*, her sail catching the wind to no apparent purpose.

A floating bar is now tied to that ring; and just across the road is a large stone memorial to Samuel Plimsoll, who, so the inscription tells us, 'fought for seaman's rights.'

I suffered rather a lot from staying in digs. I had just one room, with a bed. In the rest of the house the large, dominating landlady henpecked her husband round the place. There was of course no television, and I couldn't afford a wireless set. Evening after evening I had nowhere to go. There were evenings when the loneliness could be cut with a knife - where the silence spoke hour after hour of my failures, and seemed only to echo my sense of non-acceptance by the world. My fondest memory is of the door opening from the landing, and in coming a tray, for the fourth night running, of pilchards in tomato sauce. Trying to "build me up".

Finances were bad. I bought an old Morris tourer for £5 and spent the winter finding out why even that was a swindle. The youngsters who drive about in flash sports cars these days are lucky not to have to cope with the problems we had: either petrol to the coast, or food - but not enough cash for both; throwing meths into the clutch, just before a hill, to stop it slipping halfway up; and catching all the little crabs that scuttled round their home in the leaky bilges, guessing which bits were poisonous and boiling the rest in a bucket (supper!). I remember how mud and tide ruled those weekends. Sitting cross-legged on the floor, my hair touched the cabin-top. But I was proud and self-possessed, as if I had inherited the lore of boats and of the creeks, and the secrets of the sea-wall and the lonely, wild saltings.

Dad had got hold of one of those new caravans that you could tow behind a car. He kept it in a field near the Surrey Hills, and he knew the footpaths for miles around. Walking with him one day through the woods we came upon a young damsel riding alone on horseback. Dark-eyed and beautiful, she had to me a remote fairytale look, yet a voluptuous easy presence that quite bewitched me. I had fallen in love for the first time; and for a long, lasting time.

I suspect that Dad must have got some kind of genuine but vicarious pleasure out of hinting that there might perhaps be a stable somewhere around. With a courage that surprised myself I booked my first ride for the next day.

I am not sure what happened to *Annabelle* - I somehow hope that it was the end of the season and that I didn't just abandon her. Whatever the case, I was soon working in the stables at the weekends - sharing late fry-ups in the tackroom-cum-office. To my romantic mind this setting was charged with the direct opposite of the deadly meaningless suburban nothingness of my bed-sitter evenings.

The Two Himalayas

Here I was accepted sweetly by the young girl with flashing dark eyes, who climbed a tree to read a book, whose family could be traced back to a national hero of Nelson's days, whose great family house, stabled and towered, and with twenty-two bedrooms, set in a hillside of woodlands, seemed dizzily pregnant with romance.

I took her to the cinema in the local market town, confused by the contrast between this - all I had to offer - and the charismatic life-sized portrait of her famous ancestor (William Pitt) that hung opposite me in the huge old dining-room. Maybe, driving her to the cinema, I was able to find my tongue. I cannot now remember.

I do remember that film, *Smiling Through*, about a poor young fellow, all alone in the country on a push-bike, who finds himself bemused by a dark-eyed girl on horseback - and the romance ending in an embarrassing mirror-image of my own secret longings and clanging poverty. Whether or not she appreciated the irony, I will never know, she was as tongue-tied as I was.

I stored my feelings up until one moonlit night in the coach-house stable doorway I tried to kiss her, but she ran away. It wasn't until years later she told me that, although country-bred, she thought that kissing might have given her a baby.

Some year or so later, in spite of my continued quiet devotion, I was to learn that the young proprietor of the stables had felt the same way about her. I think it was my first sight of a woman's breast when I was allowed to watch her feed his baby. She had grown up, free enough to try and please me that way. I watched, and still loved her; and went back to my room.

It was to be a long time before the fading from my dreams of the woodland pool where I would somehow encounter her, wildly unprepared in the tempting water; a long time before that special nakedness passed away from my dreams.

That was pure and lovely; but I longed in vain.

Bill, at the office, referred me to a Mrs 'H' for new digs. She had two daughters, he said, "who don't look as if they are fed on pilchards and tomato sauce every evening."

A bit scared of my own rather unclear motives, I went to see Mrs 'H'. A fantastic Bentley Sports drove off with the elder girl (a real cracker!), leaving me talking with the mother. I think I was so green that I didn't know that broad-mindedness existed, or even what it was. Anyway, Mrs 'H' had just that. When I think back, she must have been exceptionally gifted in that department for those days, because she caught the Margate train that evening, leaving me in the care of Peggy, aged seventeen.

I can't remember the mother's return visits, if there were any. But I do remember that robust and very comely young landlady whose friendliness flowed into my life. We were constant companions. Those who ever noticed such things must have thought that we were newly married; but the little fly on the wall knew differently.

I paid for the coal, and we both warmed ourselves; but we didn't warm one another. The 'physical' we both wanted never got started - as if we were both magnetized, but hindered by a final barrier. Honour and chastity were poetically mixed in my mind. I don't think I ever formulated the idea of disturbing an inch of skirt anyway, because nervousness would have slapped my mind like an old maiden aunt. I used to long for the fireside intimacy to melt the barrier between our bodies and bring that bliss which never came; but never, ever, did I even question the sanctity of her bedroom (I doubt if I knew where it was in the big old house).

The Two Himalayas

I wove romantic abstractions as we gazed into the fire - a quarter of an inch apart. And longed, I think, for a little music to dissolve her clothes. I almost thought it would do that for me, if I'd had any music.

She would hear me arriving back very late on Sunday evenings, and get up in her nightie and dressing-gown, and make me a Welsh rabbit. How, with such provocation, could I not have slid and worshipped at the same time. How poetically and nervously blind I was.

I paid for my faint heart by seeing her one morning sitting on the milkman's knee, a bottle of whisky on the table. For the second time my heart sank without trace.

Longing years passed, and the war came and went, before there was anything remotely like another fireside moat round a young body, where I could wait a little shyly till the idealistic edge crumbled and collapsed.

I'm not sure what in fact I did learn over the years. Possibly nothing much more than predictable doubts about idealism; and I suppose I unlearnt a bit of shyness.

This process was helped and accelerated by a tour of France with twenty-three students of the Institution of Electrical Engineers. Only one of them really shared any non-electrical interests. He asked me, and I asked a taxi-driver, and we soon both found ourselves with a naked female on our laps for the first time. We looked at one another wondering who does what next, but after a drink they soon got fed up and slid off.

I recall the shattering of my middle-class respect for greying-haired and well-dressed 'gents' when one such entered the room and was immediately leapt upon by a girl who was all lovely and bare. They seemed glad to see one another again; although the back-to-front piggy-back he gave her on his shoulders, as I seem to have remembered it, meant that he could only see her belly-button. And her thighs round his ears could have made conversation a bit stunted. I was quite amazed and perhaps a little disillusioned. The old-time Twickenham pilchard digs seemed very far away.

I also had a memory of another such place, where the underlying sadness of the girls seemed to oust all the prurience from my mind. I sat with a young girl for a long time. She seemed not quite to have lost all her bloom. Ideas of rescuing her from the misery of that place took up my mind; but our coach took us away.

Foreign ways did rather shatter one's ideas. The only time I even went for a walk with a girl in those days on a lonely holiday across the Channel, she sat down quite near me in the lamplit gutter. Unused to the Continental lack of inhibition about these things, I took it as a nice, cosy, little friendly sign. Feeling, I suppose, that my long era of frustration might soon end, I strolled her out along the concrete mole towards the lighthouse. It was dark, and my arm was around her. I raised it gently above waist-level and got a hearty slap. I suppose that could still happen? Does it? I don't know. Do they? I think she was a Belgian: such endearing young charms.

And then, at last, a night in a cottage with Olive, the daughter of my landlady at the time, who I think sort of loved me but was sadly almost mesmerized into marrying my big, fat, lazy fellow-lodger.

But nearly every evening was spare and introverted, and very drawn out. There was the occasional bounce back into a kind of religious groping which really was my loneliness boiling over, but running into the sands of disenchantment. Passing my time with other youngsters was still out because of ball-games. I couldn't even play darts, which can be very embarrassing, and placed almost any Pub activities out of bounds. I stayed, secretly, night after night at the office, hanging up a dart-board and failing miserably to hit it.

The quiet Irish girl upstairs in the Richmond digs took my washing home. But there were no plunging depths of feminine mystique, so I withdrew from such happiness as we might well have found. Her nice shyness didn't deserve that.

And then there was Kathy, a nice little widow who visited my mother. (It had to be either a landlady's daughter or Mum's friend). Kathy was sweet. At twenty-eight, she was only a couple of years older than me but had a flash of white hair just above the parting. She needed love and affection, because that white streak was the result of her complete wedding-night ignorance. He must have been a nice bastard, an understanding fellow.

Reacting from the ruins of this marriage she fell in love with me, and her physical passions (paradoxically enough) got out of their psychological corner and made her dance on the doormat: "I can't wait, I can't wait..." and she would hop about, mostly up and down, as she was opening the door. She was exhausting. A nice change; but it didn't make me happy, and I left, a bit unconvinced.

Exhausting, yet I remember that blankets were a *sine qua non* to both of us. How strange that neither of us could ever have heard of those wider ideas that are scarcely questioned nowadays, at least among the post-war generation. So many little asides in casual conversation are now accepted without offence that there must be many possibilities of an unoffending, unoffended relationship. It is amazing to look back on my young manhood and then to read this, written by a well-known clergyman in a reputable national newspaper:

We want a more humane and liberated approach to sexual morality. A clergyman should not be synonymous with a 'wet blanket'. Christian Lib is a liberation of Christianity from the false teachings which have rightly repelled so many men and women of goodwill... What moral difference is there between piano lessons and petting lessons? Only this: it matters less if you don't know how to play the piano than if you don't know how to give acceptable sexual caresses by your late teens.

You are very unlikely to make someone's life a misery because you never did your five-finger exercises. But you may ruin a marriage because you never did one- or two-finger exercises. You can probably get by without speaking foreign languages, but without the gift of tongues you are likely to be a disappointing husband or wife one day... To tell lies with your body is as bad, or worse, than to tell lies in words.

I liked Kathy, and I think in a way I was kind, but I had a cruel level of expectation of unplumable depths in a woman: and, afraid of life without this, I left.

Life was so barren that an odd flicker stands out still. A camping holiday with my parents revived the almost buried idealistic, chivalrous instincts. The girl in the next tent had been terrified by the attentions of a knife-wielding camper, supposedly 'staff' of a local mental hospital, who was suffering from stress. Understandably she was desperate to get away to London. I had a reasonably good car by then - an open sports car, an old but lovely Lea-Francis - not to be ashamed of. The knight in shining armour ... except that she jumped on a train.

The Two Himalayas

That vehicle meant a lot to me: cars could be a problem. I remember getting half a crown a week from the Electricity Supply Company for the privilege of providing a push-bike on which I wearily peddled round to the houses that had decided to go over to electric lighting. I licked a red label and stuck it on the wall, usually in some nasty cupboard under the stairs, or some noisome cellar. What a thrill it was to cycle out on a winter afternoon some ten miles and find 'Key at Bloggs', and cycle there and back again. Nowadays for perhaps a slightly more senior job, you get a nice fat car allowance. And I graduated to motorized transport as soon as I could.

A succession of old, unsaleable scrap cars (usually uninsured) provided intermittent transport in those hard-up days. But one day I came across an old sports car in the corner of a garage. An Italian named Ximenes had run out of cash half-way through his payments, and the garage owner was getting fed up. I put up a very weathered old Hillman, my old £5 Morris and a 'credit note' for £5, and became the proud owner of half of this Lea-Francis. Ximenes owned the other half. It was only insured as cash came to hand, but he drove it all the same.

I was dragged very unwillingly into his rather crazy life. He had fallen in love with a girl who used to wear Post-Office red jumpers. He couldn't pass a letter-box without thinking of her, and sometimes literally frothed at the mouth when the sight of one on some street corner inspired him to flowery talk of her.

I persuaded her to come out with me one evening. He got to know and rang me to meet him. I remember having a bath to be clean for Staines Hospital, but somehow he trusted my version: that I was taking her out for *his* sake.

I suppose this catalogue of spasmodic, years-apart relationships was doomed to have an in-built, crumbling fate, consummated or not. At least they were odd highlights in a world of little else but my lonely boat, a world in which many days seemed a year - and the years were nine; which brought it to 1939.

Sometime just before this I had became so disillusioned that I joined the Territorial Army. I was very nervous about the drilling and so on, but I was fed up and felt out of things. It was a searchlight battalion, where hopefully they wouldn't take my rather unsoldierly appearance too seriously, and wouldn't form fours too persistently.

Fortunately the car gave me some kind of artificial self-confidence. It was the only one on our small camping site, and all nine of us could pile in (somehow) and get away somewhere. Cars were pretty rare - rare enough to get us a salute from the guard as three of us dodged the last night at some wet canvas HQ. A sporting sergeant crept up behind us half-way home to pour whisky in our tea in a Newmarket cafe at about 0300 hours. He didn't like last nights either.

We were called up in the war-scare at the time of Munich. The mood of the country was tense, and although not to be compared with the volunteering fever of 1914 there was no question of having to pay for petrol. One garage even gave me a free tyre. (This kind of thing had evaporated by the time war was actually declared.)

I remember us putting on gas-masks and horrid yellow capes for a gas scare. It was a nasty feeling. We peered at the Fairey Battles flying towards France, feeling a bit better for that encouragement.

That first year our chief enemy was boredom. I hadn't the right mentality for searchlights, especially in a phoney war. Boredom and indifferent health are a nasty mixture, especially when guard duties seemed to be interminably pointless. Two poor bastards were on guard out there all the time, when a quarter of an hour was often an eternity, sometimes the stars coming and going in your sleep-crazed head.

I suppose it might have been more acceptable had there been any likelihood of action, or even danger. But every night those not on guard would suddenly be woken up for the task of swinging the handle to start the stiff, cold diesel engine over by the hedge. We would crawl out of our beds - the Nissen hut fire would be either roaring or out - and we all looked as if we thought the war and detumescence were one and the same thing. It took the authorities nearly a year to work out that one and the same man could listen for planes and watch for intruders at the same time.

Our officers' enthusiasm for guard duties seemed to reflect their general mentality. In those days practice bombers flew, unbelievably, at about 120 m.p.h., and their course could be shown on a squared map by means of tiddly-winks. Rather more unbelievably we had to dust each side of every tiddly-wink each day (and sweep the hut ceiling twice a week). I don't know what the effect of this kind of activity had on the stronger of us, but it certainly caused psychological-cum-loyalty problems for me.

I was taken off this duty to spend my nights repairing broken telephone wires lying across frozen fields. The arrival of new apparatus, which was complicated in a cold and boring way, completed my depression. This all sounds very wrong-headed, but I believe that months of such uninspiring 'leadership' and no trace of excitement just killed the adrenalin that might have made my brain work in the cold.

Although I was not without friends, a little stray dog needed me most. The others seemed to have interesting lives; this camp was not where they came for spiritual comfort. One bloke was on about two women - marrying the wrong one, and so on; something about lipstick in the wrong place - which I didn't understand, and would have been almighty jealous if I had. In my experience, kisses from women were above the shirt collar.

Another was the heir to a fortune, but he liked the monotony. Complained in the same breath about his carburetter and "Bond Street models with cold bums". Awestruck, and very envious, I didn't know how anybody could equate such trials. He was mates with my friends the erks; they were untroubled by his Greek learning, his grumbling about their borborygmic habits in the confined hut at night-time. Possible, though, his really best friends were here, and disillusion in London. I think I most envied him his path to disillusion.

We were a mixed bunch; and in case you don't know, Army life was crude: except we had a fine revulsion for hypocrisy, which somehow redeemed our standards - and made us love old 'Smiler' and his plumber friend. We were all really kind of mates, but I liked these two best, and had a soft sport for Smiler in particular. Like I have, in fact, for the Archbishop of Canterbury - only different. That plumber friend taught me to fix noisy tanks in the loft with an old sock - possibly not quite Lambeth Palace.

I remember old Smiler over one weekend in particular - the Monday morning. His weekend 'relations' had been a bit of a tragedy because he first paid ten bob, then half-way down the stairs had been conned for more "or get back out". Then, finally, he had been handed a bar of soap through the curtains - not quite what he'd been led to hope for.

I know that Dunkirk was in many ways a shared national experience, but the sudden imagination-galvanizing news that the little boats were leaving clutched at every fibre of my being. The newspaper that told me about it showed a full-page picture of a girl who was sailing across. I can see that picture now.

God! the number of times I have since cursed the stupid kowtowing to 'authority' that kept me in that field, while my little converted whaler-ketch lay at Althorne on the Essex Coast.

I heard years later that others had jumped the lines and untied their boats; but some well-meaning elderly Sergeant talked 'sense' into me that night, and, easily influenced, I stayed. A special, peculiar shame . . . for which, in my book, no forgiveness can be found. Boats were few enough in those days. Later I discovered that the Admiralty had called upon any boat over thirty foot long; mine would have qualified by two feet. I could have saluted my CO and waved goodbye to the Army there and then, with the paperwork being sorted out later. Except, of course, I might not now be here to feel sick about what could have been.

As if to rub it all in, my younger brother had recently obtained a commission in the RNVR, as an electrical officer. Saluting him outside a pub in Portsmouth, I listened to him say that he'd rather exaggerated the two days he had spent on my *Annabelle*. It had been enough to impress the panel; sailing was less widespread then and had a currency of its own. Perhaps he sensed that it still went against the grain for the officers on the panel to put the uniform that meant so much to them on somebody who didn't even care about wind and tide, about picking up moorings, and about coming alongside.

The irony of the situation and jealousy rumbled within me. I had tried every official and unofficial way of wangling a transfer from the Army. The word from the Admiralty was that you had to be a professor of something which the Navy needed and the Army didn't.

I think it was the ringworm that saved me. The Army had, after weeks of washing up and sluicing the germ around, eventually put me in an isolated room, where I heard on the radio that Churchill had ordered the capture or the sinking of the French Mediterranean Fleet at Oran, to prevent its possible defection.

Professor of something? I remembered that I had matriculated in French at school. Having written letter after careful letter, I couldn't be bothered to reach for a decent pen that had rolled under the bed, and so wrote carelessly, kneeling, the paper propped on the bed, to the Secretary of the Admiralty:

> I beg to bring to your notice that I can translate technical electrical French. I suggest this may be of interest in view of the reported acquisition of units in the French Fleet.
>
> Yours obediently,
> A.F. Apps

Some days later, to my amazement, I was called to see the CO. Having first rollocked me for having written over his head, he broke into a kindly smile, said that I was to have an interview at the Admiralty, and asked what he could do to help. (How I'd like to buy him a drink, whoever he was).

I think I must have confessed to him that I could not, in fact, translate that silly technical French, but it had seemed possible that the Navy might be rather fancying a professor of that sort, etc., etc. In retrospect, it really does seem a most unlikely ploy. I think that it was just because of that that he made it his concern.

The Two Himalayas

We talked it over. He decided I had better have two days leave to find some technical dictionary or other that might rescue me. A taxi to Loughborough library finally ended my search. There it was! pages of little diagrams of electrical and other mechanical bits and pieces, with 'earth; *perte a la terre; Erde Verbingung*' and suchlike, plus three other languages, in columns alongside the original English. I soaked it up night and day. I got put on guard-duty and read it; I asked to do potato-fatigue each day, and hid it in the heap of spuds I was supposed to be peeling.

Finally, thoroughly alarmed and nervous, I went to the Admiralty for an interview. I joined five other candidates - all seemingly bristling with qualifications as electrical officers. I remembered my pre-war failure even to qualify as artificer on the RNVR's HMS *President* on the Embankment. The old-fashioned 'straight up and down' private's uniform made me look, and feel, skinny and ridiculous; it further embarrassed me by its fifty-two shining buttons.

I had worked myself up into the conviction that there would be some French-speaking electrical engineer on the panel. How naïve I must have been - and nervous; so nervous that even my hearing seemed to give out, with the result that the senior officer's opening remark, actually in plain English, shattered my frozen phobia. Suspecting some sort of paralysing French, I replied, "Excusez-moi, monsieur?"

He slowly removed a monocle from his eye, looked me up and down and repeated what he had, of course, first said: "I suppose you want to get out of the Army, don't you?" Recognizing my native language: "Yes, sir. - Yes, yes," I babbled. The rather puzzled-looking row of fearfully senior-looking officers sitting round the table facing me, found this all so ludicrous and hilarious that even I caught the infection, and my 'nerves' went strangely out of the window.

Yes, I had had a sailing boat - and a brother in the RNVR. The nature of their smiles changed, from one sort to another. Batteries? I had worked in a small battery firm, I lied, now somehow cool, relaxed and confident, feeling that they weren't really interested, that none of them really wanted to intrude electrics into a comical situation vaguely involving boats, and a brother already in anyway. The French translation bit didn't arise. I suppose that I really have to thank the chap who couldn't find the right file for my original ringwormy letter - and then the fates who pin papers together in lucky and unlucky ways; and then my own poor nerves.

Some days later, when our searchlight unit was back at base learning how to form threes (for some unknown reason), my CO received a telegram from the War Office to say that 2052157 Sapper F.A. Apps was hereby released from his military obligations. Words that can still ring in my ears . . . .

I am told that I was the first to get out of the Army into a Naval commission. It may be true - first one below the rank of officer, I suspect. Be that as it may, when the 'Fall-in' sounded, the reality dawned on me; I stood lounging in the NAAFI doorway, watching my suffering colleagues, and flaunting a large mug of tea. All the hopes for the future and the end to past frustrations seemed to waft from that lovely brew.

A few days later I was in the train to HMS *Vernon* at Portsmouth, still in soldier's outfit - but with a naval uniform hidden in my case - swapping moans (the very sweetest, to me) with a carriage full of soldiers. I am sorry, Mr Army, but you do, or did, ask for these sentiments. "You (other ranks) *will* do this; so and so *will* report." I even remember (well after the war), a huge notice hung obscenely in a lovely wood

The Two Himalayas

that I passed every day on my way to work. It read 'Trees Will Not Be Damaged'. One morning I got out my starting-handle, went into the wood, and smashed it down, and sent it in a big box to the War Office, requesting them not to something or other well address the public in that way. Of course, I often feel admiration for the Army - provided I'm not caught up in it, I suppose.

I felt somewhat alone in the streets of Porstmouth, and also slightly apprehensive, as I reckon most of us are when joining something rooted in history. A little subdued, and rightly so; more worried about procedures than about what the war might have in store for me.

Although so near to old HMS *Victory*, I am sure that I never gave a thought to such sentiments as 'England Expects'. But if I *had* any confidence it was, I suppose, subconsciously rooted in something like that, or in some lump of tradition that was stuck inside me.

The training at HMS *Vernon* was all about magnetic mines; but the thing that I can now recall most clearly was that junior officers all read *The Times*, or *The Daily Telegraph*, while, incongruously, the top senior end of the breakfast table was dignified by the *Mirror* and its tabloid equivalents of the time.

Many of my colleagues got posted to the East Coast to help in the struggle to counter magnetic mines. I was sent, however, to Londonderry, where there were quite different problems, and where I was to fall foul of a different sort of mine - departmental petty-mindedness. It blew me, eventually, to India.

Despite the disappointments, I never regretted this Londonderry posting. Suddenly I was where I wanted to be - involved with destroyers in one of the great dramas of their colourful history. More than a third of the Western Approaches escort forces were stationed there, it being our nearest port to the USA, and the Battle of the Atlantic was one for our survival. And to think that, but for my 'expertise' in French, I might simply have read about it in some field, bored, and with no heart left for anything . . .

There was only one scare about mines all the time I was at Londonderry. With a convoy losing twenty out of its thirty-four vessels, Captain Ruck-Keene wouldn't hear about anyone electrical sitting about over there when he needed his destroyers and corvettes and little armed escort trawlers to sail free from those pestering strings of electrical defects which otherwise kept them tied up in harbour. And, by God! nobody could have been more thrilled and more willing than I.

They mostly spent some ten days at sea and about three in port. I knew that their bunks were seldom dry in winter. Nobody spoke much, if ever, about the sinkings.

I hold it to be a quite remarkable and very significant thing that a senior officer, Commodore of an exhausted convoy that had probably only just docked, should stand up when I, a young RNVR officer and very green, entered his wardroom. Perhaps it was the resilient strength of a wonderful old tradition, something very near the true core of their personal discipline, and very near one of the real reasons why they won't be cowed in war. Whatever it is, it affects - infects - the crew. So that, even if they, the hands, have metaphorically to clean their tiddly-winks, they know what is behind it.

I don't know what happened to my shyness. I suppose I felt in my element - apart (often) from the electrics. Lots of wires can quite easily baffle my sort of brain, and

The Two Himalayas

I decided from the first not to bluff. I don't think the 'Torpedo' (electrical) petty-officers would have worn too much cockiness. A touch of humour was more important, and they helped me in a thousand generous little ways.

It may seem strange to recall, in these troubled days, that the local Londonderry people were quite fantastically helpful. There was a Mr Kennedy, who ran I believe the only local electrical wiring firm. He found dozens and dozens of completely willing - if a little inexperienced - 'wiremen'.

It amazed me the number of light bulbs that went missing; but it also amazed me how hard and persistenly everybody worked. The only real indifference I remember seemed to come from the local City Council which forgot to cancel some ridiculous rule that Town Hall dances should finish at 10.30 p.m. on a Saturday night, leaving the sailors nowhere to go. Captain Ruck-Keene, a widely respected officer, believed that the local population thought and felt otherwise. He posted two armed marine sentries at the doorway, with a typed notice that they were to gently resist any member of the city council who tried too hard to insist on the 10.30 rule. He wasn't far out. We gathered that the daughters canvassed their councillor daddies and it got reversed more democratically. Everybody was all smiles, as far as I know.

We think of Londonderry in the context of the unfortunate religious troubles of recent years. I encountered such ill-feeling on only one occasion, when I hired a Roman Catholic taxi for a Protestant wedding, and so got badly stung. On the other hand, I learnt that the local RC priest had a Protestant maid, and the Protestant priest an RC girl; a particularly touching piece of co-operation based rather lovingly on the Pastors' appetites. The timing of the two religious services left each girl free to cook a Sunday lunch - a religiously tolerable arrangement, I suppose.

It would never have done, though to have forgotten to "fall out the Roman Catholics" at the Sunday church parade. I waited - fortunately perhaps - until I had got drafted to the south coast before I managed that one. Down there you could get away with it; here it was different.

Ships entering the River Foyle to reach the Londonderry base had the neutral Free State one hundred yards away on the starboard side, and visiting US ships sometimes decided to anchor and think about it before venturing further up this puzzling river. Until they had been made ready for sea again in two or three double-quick times, they were wary and suspicious about sabotage. My best Irish charge-hand wasn't far off being arrested, suspected of trying to delay the convoy by damaging the gunnery control circuit. Whilst the boiler-suited commodore crawled unthinkably over the filthy boiler to have a look himself, I twisted the wires to confuse the issue - sure of nothing but the man's loyalty. With my heart in my mouth and my bluff nearly called I watched the commodore crawl further along the wrong wire ... and then suddenly give up, leaving the scene of my reckless intervention. That was dicing with my integrity. I suppose loyalty begets loyalty, but my deceit would hardly have been understood officially. As it turned out, the Navy had thrown a complete US Navy mast overboard in a lull between two storms while bringing over one of the fifty old destroyers that Churchill had swapped for island bases. It was, as usual, top-heavy with

The Two Himalayas

Yankee gadgets (so irreparable that they nearly lost the war at times). The wiring had got ripped: hence the 'cut wires'.

Generally speaking, however, there was a charming 'Alice-in-Wonderland' side to being in that part of Ireland; unlike Belfast with its Union demarcation about who did what that tied up some of the smaller escort ships. I well remember having to wait a day for a fitter to come and unscrew a rather big nut - while unforgettably, the sinkings went on.

There were advantages. Free State goats were ferried across to be disguised by London hotel kitchens. Free State butter didn't all melt in your pocket in the slow, hot bus across the Border. As for the smuggling down by the estuary, I can't imagine why the Navy even bothered to try and stop it. It had always gone on, and always will go on.

The skipper of the diesel patrol boat, Dodd Dodson, was a very colourful and independent personality, and I'm doubtful whether anyone was really convinced about his anti-smuggling role along that River. Perhaps that's why he was chosen? At any rate, I got to know him well, and admired his strength and ability. He and I used to go off along the coast after washed-up mines, and had a bit of a special relationship. So that when something went wrong with his diesel, and it was almost glowing, instead of coming alongside he got me rowed out (not very RN) to his unusual midstream mooring. Well he might. Months later, I watched his strong arms and the beginnings of a smile on his weather-beaten trawler-man's face as he told me that when I was aboard I had nearly been sitting on a woman. He had her hidden under a heap of blankets, and she had come within inches of jogging my cup of tea. They drifted downstream that same night . . . .

I felt rather honoured to be trusted with the secret of this naval stowaway, and intrigued as to whether there could be another character like Dodson anywhere else, or a long, rather bewildering estuary to make another such story possible.

After the War Dodson got into the news by disappearing with a trawler, the *Girl Pat*, to the Sargasso Sea, instead of fishing colder waters for the owner's profit. I don't suppose the owners and authorities were much impressed, but I lapped it up with boyish delight. Wherever he is, I wish him well.

Magnetic mines were swept by discharging rows of batteries, once every hundred yards or so, through the coil of wires wrapped round a towed barge. The thousands of amps made a magnetic field stretching far enough in front of the towing vessel to explode the mine before it got too near. Perhaps that's not a very full explanation, but it doesn't matter.

One day, a trawler full of batteries was moored right alongside its coil-barge when I closed the big main switch to see how it worked. It worked all right, but it magnetized the iron-built trawler, which itself became a magnet: a sort of first-year physics effect that made its compass go all funny. They had to build an eighteen-foot high wooden tower on the foredeck, and sit a man up there with a compass so that it could find its way to Belfast. I almost think I gained in prestige by this demonstration of anti-mine phenomena, at least among the humourists.

Another intervention achieved even more spectacular results. I had tried to make it not too obvious that the West Ham Technical College evening classes had never included taking submarine main motors to pieces, or what to do about nasty noises in general. So that when I found myself down in the engine room of a rather ancient training submarine, where I was expected to detect some fault or other, I was paralysed by the size of the obstinate monster and needed time to think. The fast-revving always overloaded, machinery of the newer ships scared me enough, and I was forever dreading some revolving crunch or other; but this old relic was going to guarantee a real struggle with my rusting theory.

In trying to hide my embarrassing ignorance, I must have made some casual remark that got taken seriously. By next evening the captain had organized himself into the little repair dock, unbolted the forward deck-plating, and got the huge 900 horsepower motor hanging in the air from the dockyard crane, leaving the sides of his vessel only about five or six inches out of the water. With very mixed feelings I realized that this meant sending the motor to Belfast. What if there was nothing wrong with it? There might very well not be. I shuddered, but the war pressed on, and I heard no more. I daresay my stock with the wags increased significantly.

Opportunites for charming little worries of this sort are rarer in peacetime, where you can usually organize a meeting to cover your tracks. I suppose this is typical of "how we won the war, mate!" Or could it in some ways by *why* we won it?

But life can never be perfect. I was still plagued by hangovers from my early days of hiding from the worlds of sport and social mixing, and once again solace seemed to come via the landlady's kitchen. Or kitchens. The first affair was to end, in fact, rather painfully and I would need another kitchen to get over it.

I had been to Belfast, where the landlady's daughter had anorexia nervosa so badly that she was hardly more than a skeleton. I found her most unattractive, but as her rather harrowing story and little remaining bit of personality unfolded, I found myself getting involved emotionally. I held her hand and showed her that I cared, which I truly did. Some kind of answering light came and flickered in her dulled eyes. And as meaning came back to her life, of course flesh came back on to her body. When I was posted to Londonderry, she decided to join the WRNS and follow me there. She must have wanted to be pressed into love, as Olive and Kathy had been. On reflection, it could only have been that my old complicated idealism had come back to keep me from touching her, because the manner in which our mutual feelings had grown (out of living caringly in the same house) could have left just no room for nervous shyness.

It would be a sorry world without idealism, but it can be a sad one without a little balance. Reality has its virtues as well.

One day she suddenly scrambled down a cliff inviting me to swim in the nude together. I had no prudish reactions, just utter shame and frustration because a gorged stomach kept me at the top of the cliff. I am still annoyed with myself over that vexing memory. Something more for the humourists. This unheard-of nude bathing was in fact very close to my secret ideals of mutual trust; the frustration came mostly from missing what could have been the idyllic handling of a quite fantastic trusting experience, for those days, at least. I am sure I would have been half acting out my dreams. Olive and Kathy were, I suppose, too simple.

This strange mixture of idealism and lack of prudery, and of sex sublimated by trust, dominated me still; the girls of my dreams and daydreams loved those sublimated ideas and found them erotic and didn't mind. But this real person (apparently also more simple), after a night in a hotel bed and philosophizing hours in a far-too-thoughtful contact (and no more than that) left me and those beaches for a colonial.

She had started getting pretty and promoted. I'd felt proud that her gluteimaximi left an ebbing tide of Naval absentmindedness, distracting the clerical ratings in her office. But the cleavage between them as she warmed her thing in front of my fire, only made me want to turn her round. To turn her round but with almost indecent haste to seek upwards to her eyes - because ideas lived in eyes that took the brunt of passion.

I must have been a bore. Poor thing.

I remember vaguely this general idea of intimate non-consummation attracting another casual but more romantic girl; but the same rather intense experiment produced the same result. She also wanted to go to sleep with sparkling eyes. I wanted to stay balanced but breathing, eyeballs on the edge of sex, as if the greatest thrill to both of us was in keeping some unclothed vow, keeping her final chastity on some prolonged spiritual hilltop, never falling over the indulgent cliff.

I don't think there's a word for this: nor for untying that ensuing mental knot. It must be simpler to be a prude.

I was to compare all this, many years later, with the untortuous experiences of a friend who had lived about the same time in the remoter reaches of the Norfolk Broads. He was young then, and had rowed and swum amongst the reeds from childhood. Boy-girl relationships were at a very shy young stage, but one day he just stripped off and dived in. He remembered how she said, "Oh look! you've got no clothes on. How super! Can I come in naked too?"

He was to find hardly any other reaction through several rather philandering years. How was it, I wonder, that girl after girl should take so avidly to this, in an age when there was such a crippling Victorian set up; when today's sleeping around was uncommon and so would not have debased the adventurous coinage of such an experience?

That way, he said, he first learned that the hair on a girl's head blossomed rather cutely elsewhere. I imagine that I must have been about five troublesome years older than him when that mystery was first solved for me. Perhaps it was just the Norfolk air? Once out on these remote Broads, maybe I would have jumped in too. Perhaps landladies' daughters as well?

Maybe - no doubt because of the hassles and traumas of the emotional prison of my youth - I could have a biased, wishful-thinking approach to this matter. But it does genuinely concern me that 'protection' of our children in fact achieves the reverse. I get a vicarious pleasure in vaguely imagining the falling away of the accretions fostered by our taboos. I have tried to fuse it into the deeper hope that runs through this book. To me, there is a precious connection.

The word for this desired state is not quite 'freedom'. There can be too much of that. Perhaps it is something like a secretly virginal uninhibition; but in a very precious sense untwisted, possibly a little sacred. For the time being, though, I know that some readers will have doubts, and I sincerely ask those who cannot shake them off to bear with my genuine hopes.

The Two Himalayas

They are hopes, these little coves, for a purging of the effects of pent-up complications, bringing them to the surface as they run in and out of the waves. I suppose a sort of catharsis can happen, which buries the past and leaves it in the sea. For the young, those who were lucky enought to start life in the new way and be brought up in such a tradition, all that wouldn't have happened. And still it might be freedom, with lots of the nastiness taken out of it, the nastiness that hasn't got a chance in the waves: sex being defused, diffused, and degiggled because it is depolarised from the sexual organs and focused on the whole being, those parts that rivet the attention in a strip-tease affair no longer being a magnet for the eyes?

I said that I needed another kitchen to get over the anorexia nervosa affair. This was not far away - downstairs in my Londonderry digs; the landlady's daughter, needless to say. Slowly the magic formula of proximity and youth did its work. There was no heavy, searching, soul-mate stuff and no idealistic theories, because there wasn't any drama for them to latch on to; but we developed the knack of a little cuddle on Sunday morning before the house was astir. This time my thinking was concentrated more on how not to get caught, as I crept up the creaking stairs to her attic room, a cup of tea balanced in my hand. I practised creeping past the door where her parents slept. I believe I could almost remember to this day which end of which stair would give me away.

Conventional religion still had some influence, enough to set my conscience a poser one Sunday when I found her still in bed, but reading the Bible. I sensed that to slink out of the room again would have stamped our little pleasuring as incompatible with our respective religious sensitivities. To interrupt, get the book laid aside, and apologize afterwards seemed the same, only worse, for her. False values either way. By some kind of telepathy we didn't want drama. Nothing was said. *Quite* possibly nothing was read. But it seemed almost as if the true message of the Bible didn't have any patience with me and my past (or with some of its old interpreters, come to that), because I don't think it minded; because she, seemingly unperturbed, never lost her page.

I somehow delighted in the no bad conscience and the no frustration. It was an honest little episode, even if with tell-tale little signs of youthful ways. It seems as if I could at least sometimes have been aware of an emotionally disentangled penis - when the profile of my dreams was quiet and low.

I was pretty unsophisticated too about the 'border question'. I think most of us were. When I fancied trying to shoot duck (something I couldn't bring myself to do now) the local police didn't want a licence. They just handed me some kind of sporting rifle which had 'For God and Ulster' elaborately engraved on the wooden stock. It meant very little to me. After all, the locals worked more than hard; the border was just an amusing diversion - especially around Christmas time. Contacting the pig-smuggling fraternity was fun, but a tin bath full of pieces of pig left at the bottom of the garden was a little embarrassing. Hiring a horse, and riding back after dark with a turkey in a sack, was better. When you saw the glow of a cigarette being waved in circles some way off to the right of the border post, you knew it was the guard enjoying your pre-arranged fag.

The Two Himalayas

A more serious diversion came my way because Captain 'D' (as Captain Ruck Keene was known - 'D' for destroyers) discovered the word 'fuse' in the second paragraph of his top secret Admiralty instructions about unexploded bombs. 'Fuse' meant electric, and 'electric' meant *me*. So he handed the papers to me to look at; I suppose I must have been rather busy, and put them away in a special place and, as commonly happens with important things which we put away in special places at home, I forgot where this was.

Belfast's first air-raid some time later initiated two days of worried and very furtive searching, hoping and praying that the Captain wouldn't ask for them. I found the prospect of being made to look such an idiot so alarming that I was quite relieved to be asked to go to help out at Belfast.

The bomb-disposal officer there had developed eye-trouble - in fact he had almost gone blind - so I was met by his petty officer. We all three then discussed the situation; I took the officer to an optician, who figured that it could easily be the result of the continuous tension of his job, and in which case he should be all right again in a day or two.

As you can imagine, the petty officer was a grand fellow, and together, plus a copy of those elusive instructions, we did what we could. The only really nasty bit was trying to do something about a bomb that had buried itself right alongside the main hydraulic compressor, from which all the heavier machinery in Harland & Wolff's shipyard obtained its power. The dockyard - or part of it - was built on soft land, reclaimed by means of driven-in wooden piles. The bomb was about six feet long and was buried head-first, deep down in the soft yielding muck, thankfully not quite vertical. If it was ticking it would go off: if you moved it it might go off. You might get a clue from the hieroglyphics printed on the round sealing plate at the bottom of the hole; or again, you might not. I suppose they must have held my feet or tied something round them and lowered me head-first down into the space beside the thing, which appeared all right as it seemed not to be ticking. Somebody handed me down a doctor's stethoscope. Couldn't figure much out from that - mostly gurgly noises, I thought. And what a shame to have to use my nice clean hanky trying to clean and read whatever was on that round little plate. I still remember thinking that, but cannot remember whether we eventually left the bomb there, or tried to dig it out - or had more than a pint of beer. War does funny things to you! I heard someone mutter something: but to me it was hardly as bad as that ball coming towards me on Saturday in the park.

Our only other problem was a clean round hole that had suddenly appeared in the concrete forecourt of the offices of the Rear Admiral, Northern Ireland. I didn't like this one bit. Neither of us could make out any ticking when we stuck something down the hole, but there was, on the other hand, just no explanation of how the hole had got there.

The admiral grudgingly moved his two or three hundred staff out of the building (as per paragraph x, clause y, of Standing Orders). Directly the commotion was over, and the place clear, we started digging. There was no bomb. I began to imagine the admiral's comments reaching Londonderry, and my name being associated with those lost papers. The very thought gave me the shivers. An idea suddenly struck me. In the back garden of the house where the unfortunate man was gradually getting his sight back, there was a big first world war Naval shell, stuck on a sawn-off tree-trunk.

The Two Himalayas

The landlady was good about it. She could (I hoped) have it back. All the office staff were warned that a very tricky operation was imminent. We drove on to the wide expanse of concrete. I got hold of a cat and walked it to safety. Then the petty officer and I slid the old shell into the hole from under a sack. We then went through the rigmarole of digging it out again. Fortunately no one was near enough to spot the rather unlikely bronze colour of the 'bomb' as we humped it very gingerly back on the lorry.

I had the feeling that the whole thing was quite transparent. But driving off as fast as I could, not knowing which gear was which, I knocked down a big chunk of the fancy brickwork archway over the main Dockyard entrance. Nobody seemed to mind.

I regretted sorely that I dared not breathe a word of all this, because it would have had to leak upwards; yet a yarn like that could have kept me in free drinks for quite a while. It was worth more than one furtive beer.

A different sort of unexploded bomb dilemma arose some time later while I was on leave in London. The new telephonist at the office where I used to work was a quite ravishing creature, with beautiful golden hair. The uniform did the most impossible things for me - I could come home to her flat in Chelsea. I drove there, unbelievingly, with her in a Taxi, only to find 'Unexploded Bomb' chalked on a bit of board outside the block of flats. We looked at one another; she would chance it, she said. More enraptured than ever, and in no frame of mind to let the Navy down, I walked with her along the corridor. Her room door was half open - my heart sank - inside were seven or eight people trying to get a bit of sleep on the floor. This apparently was, after all, the safe end. They had come from the bombed end of the building. I was quite shattered. But what a fantastic piece of hospitality. I really had the feeling that she was ready to take that sort of risk because she thought that deep down the Navy was worth risking herself for. She was too attractive for what, in those days, would have been a cynical interpretation.

Ireland fascinated me, except for Dublin and the South, whose people were a bit unsure of their attitude. Whilst in the South, on leave, I'd got the local police all excited as being a suspected German parachutist. So I readily took the advice of the gentleman who'd put me up. He simply told them to calm down, and me to catch a train to the more friendly Donegal.

I remember that holiday well. I hired a horse and promised either to sell it or return it at the end of my leave. At the first village they put it in the shafts of a cart, which it kicked at viciously, rather deflating any idea of my wandering off and just selling it when I got fed up. I had a lot of difficulty getting it to go uphill, but eventually found myself staying in a deer-warden's lodge near a lake. A castle-like building, on an island strutted over by peacocks, completed this lovely scene. After a day out with the warden, and having heard with disgust that it was all owned by an American baked-bean merchant, I carried on with this perverse uphill business until at last we dropped down into a straggling village where an old lady in a gnarled old cottage made me welcome. It nestled by the water under a strangely-shaped mountain. I discovered an overgrown sign saying 'Martha's Glen' hanging, just legible, in the porch but had ominously heard only 'Martyr's Glen' in their local brogue.

In those days it was very wild, and it felt more than remote. I fixed up a spare stable for my horse, but found myself expected to know the cure for the mare in the next pen.

The Two Himalayas

The poor thing was in pain, unable to pass water. I felt that their treatment was a bit out: they were rubbing Holy water on its forelock, and gin on its thing. Anyway, one minute I was standing there; the next my mare must have laid me out - I still have the scar on my forehead. I was told later that she was in foal. I got her sent back to the man who had foisted her off on to me.

So much for my knowledge of horses! But, as you may be now be able to guess, I fell in love with the young Southern Irish lass who cooked for me in that cottage. Her old stepmother would go off to bed, and leave us baking griddle-cakes over the peat fire. This may, by now, sound all too familiar, but I really felt this time that we got to love one another. We seemed to fit perfectly in a very rare way, very new to me.

As I recovered from that mare's kick, all sorts of difficulties began to arise. It got me down that we were not allowed by local opinion even to go for a stroll together; that the local Roman Catholic priest dominated the village, even insisting on how much these poor people should put in the church collection (a lot).

I felt that I was more than prepared to try and sort all that kind of thing out, but I suppose that the pressures were on her; because one night (when we *did* manage to escape for a walk) she suddenly disappeared into the darkness and drizzle. I feared that she might well be in some kind of state, and plunge on, God knows where, careless of bog and direction, wet and cold. The thought of a young girl alone out there in the dark, in that peaty, swampy wilderness appalled me. She did return unharmed, some time later, but it shook me up.

Somebody must have begun to see some commonsense though, because we managed to go out one day together to Letterkenny, the local market town. I was thrilled to bits to be able to treat her to a simple meal in a small hotel, but was a little distressed to find that poor Molly didn't know which hand held the knife and which the fork. In spite of this, she was, I am sure, a natural lady, by instinct if not by education.

I was considerably more disturbed to find, however, that a peep into her bedroom revealed the most awfully off-white bed linen. I lay awake that night plotting to send her to some kind of covent, hopefully persuading the Abbess to see that she was taught these things in the kindest way possible; if necessary in return for a fat contribution to the Abbey funds.

I hung on to this idea - I was in love, and so I am sure was she. I really think I would have done that. I learnt later, to my shame, of an Army officer who had. But one, to me, almost insuperable barrier suddenly arose a day or two before I was to report back, cured of my horse-kick, to Londonderry.

We were out walking, quietly in love, the bog incident just a memory we could smile at, and for the first time I ventured to stop and not be too puritanically behaved. When, next day, she said that she had gone to confession over it, I felt that I could not go on. To be regulated by priest-dominated views of that sort seemed to have a terrible significance to me. I think that some kind of spiritual pride as to my ability to judge right and wrong was at the bottom of it. Perhaps the Army chap was more human. I prided myself on my tolerance, but I was hardly tolerant towards her, making no allowance for the narrow community in which she had always lived; never giving her any sort of credit for any longing of her own towards freedom, and away from that view of pleasure-as-evil that might have been fretting her.

The Two Himalayas

Bearing in mind the narrow, crippling village life she had to settle back into, buried away from everything she may have dreamt about, I can hardly claim to have left happiness behind me. At the time, my sorrow was more for myself than for her. It should have been the other way round. I could often have got right up and gone across to Donegal, to see what light was in those eyes; to see if they were sad, buried in drudgery, or whether someone had healed them and brought back their depth of sparkle. Or whether she went away, chasing some shadow of the happiness we had stirred in one another.

I was to find that out; and I should like to think it was as much from caring as from nostalgia. I was to find that out too.

Any conviction about my search for a soul-mate must seem to be getting a bit frayed at the edges. Yet I do think, looking back, that there is a genuine place for caring in a delicate and absorbing way for more than one partner. Maybe there is room for that rather kindly old song:

At seventeen, he falls in love quite madly with eyes of tender blue.
At twenty-four, he gets it rather badly with eyes of a different hue ...
When he fancies he is past love it is then he meets his last love,
And he loves her as he's never loved before.

Genuine room for that streak of truth - which reverberated through the music-halls of our grandfathers' days - so that everybody loved it.

But it isn't always funny at the time.

Oh dear! I have watched a close relative of mine fall in love with, and happily marry, the first girl he ever spoke to. How very much more simple.

The base at Londonderry gradually expanded because convoys could be protected for an extra one hundred and fifty miles out into the Atlantic, as compared with Liverpool. An Engineer, Rear Admiral arrived. Never heard of one? No more had I. A delightful man, he gave me a freer hand than ever. But strains and stresses between the old long-established Engineer Branch and the newly-independent, fast-growing Electrical Branch produced an Electrical Commander RN - a very rare bird - who of course took over the job I wasn't supposed to have been doing anyway.

Our old pleasant relationships were replaced by official relationships, and I was recalled.

Ironically, that old pleasant relationship had been based on a row. One day I had banged my fist on the senior engineer's desk and said "I wasn't interested in his bloody bits of paper!". After which, instead of having me up before the Old Man, he had been as nice as pie. And that charitableness seemed to justify - more than justify - co-operation.

But I was recalled, and almost put into quarantine at some God-forsaken little harbour on the north-west coast, which I won't name in case I upset the local golf club or environmental officer. The bleakness of the long concrete waterfront reflected my thoughts well. I don't know whether washing-up and the other chores of an Irish village cottage were as unlovable and uninspiring to Molly as the three little landing-craft were to me. Tied up in the desolate harbour, their grotty little starter-motors were my biggest responsibility. It hardly fooled my sense of excitement.

The Two Himalayas

I had been posted to the 'Second Front'.

I longed, perhaps rather naïvely, to get back to the palpable if casual atmosphere of the destroyers: tied up, pregnant but subdued, alongside the quay at Derry. I even longed for the troublesome but salty old trawlers, and the rather tubby looking, new-fangled corvettes.

Feeling at home with *them*, I had once scrounged a lift on a destroyer down to Avonmouth. An unlit floating dock 'somewhere in the swept channel' worried me, but it didn't seem to slow them down. Was that risky? I didn't know. I don't think they had radar - they were just so desperate for leave. I learnt to take the captain's comments about the navigational abilities of some of his flock with a pinch of salt.

Touched by nostalgia, I thought I'd look up my first lady-love, who had moved from the Surrey hills and was now living near there. Sitting on a log next evening with desire unforgettably hammering at me in the darkness (me again), I promised to try and save her favourite horse from some destruction order. Those of you who are old enough will remember the countless regulations limiting the size of parcels and the like on railways in wartime. Very unpatriotically the GWR had neglected to ban the use of horse-boxes. So - very unpatriotically - Sammy and I were despatched to Guildford, where he was struck with fear by the overhead coal conveyor at the power station, and bolted with me up the cobbles into the town. A patch of green quietened him, and he took me some fourteen very memorable Surrey miles over heaths and fields to the old stables. Some time later I got him down to Hayling Island, where the landing-craft malaise returned to afflict me (perhaps rather illogically) at every creek.

As far as I know, Sammy carried on giving rides to friends when I was sent abroad to India. The duty watch used to be turned out to catch him for a pretty Wren - or on Sundays, there were the Roman Catholics 'fell-out' from Church Parade . . . .

Illogically again, I felt I'd been 'fell out' from the Navy I'd loved.

3

Having half-neglected my minesweepers, I was stuck permanently with those landing-craft - the original small floating boxes with a motor that were eventually to establish the Second Front in France.

I am a bit short of inspiration regarding my contribution here. Not very commendable, to say the least, but fiddling with their dynamoes and starters was a job nobody wanted to know much about.

Frankly, I hated everything about them. And having heard a Rear-Admiral talk of my promotion - presumably into and not out of the drama of the time - and then, instead, been put at the bottom of the class by some closed mind at the Admiralty, I'm afraid I was a little unresponsive to their charms. As I gazed down at them I felt that vague thoughts of injustice went quite well with these cold, crewless, quite characterless things. Their invasion potential - the freeing of oppressed peoples - hardly ever occurred to me.

Odds and ends of postings followed. I appreciated the liveliness and the warm generosity of the US Navy at some Loch in Scotland; but I did think their over-gadgetted automatic laundry and kitchens rather too complicated for our beseiged wartime state. I remember nothing specific - absolutely nothing - about doing anything to any landing craft, but I suppose I must have. I do remember having to sleep in a haunted room in a commandeered empty caste. Three Naval sentries had fainted while on guard duty in the haunted East wing . . . something about a 'White Lady' . . . and I myself woke up with a white lady shape hanging over my bed . . . I froze with horror. Perhaps it was just mental?

I think I further annoyed the Admiralty by asking for too many men. Perhaps it wasn't surprising that I found myself eventually aboard a troopship bound for a job training some unit for the Royal Indian Navy.

And so, at last, we come to India, and my strength and discovery and weakness and sickness there - though maybe the sickness started before I arrived there. I felt terribly weak whilst we were lying off Sierra Leone, but thought that everybody else felt the same, and so said nothing. I am now pretty sure I caught something there that followed me round for a long, long time. Some time later I heard that the Depot ship there couldn't put to sea, and divers found that she was aground on empty gin bottles. I shouldn't be all that surprised. Freetown is a sweltering place; it has the distinction of having expanded the Navy's vocabulary into hyphenated place-names that are unlikely to be printed on any map or uttered in polite company.

The Two Himalayas

The Senior Officer on the *Monarch of Bermuda* was an RN Commander who had been brought back from retirement. He called me to his little cabin and told me that I was to be 'Captain of the Foretop'. I muttered something, saluted, and quickly sought the opinion of my one fellow naval officer as to what it could possibly mean. He, RNVR of course, was equally at a loss. I felt that the three thousand pairs of eyes below decks would be watching me if I asked any one of them how one was to perform that mysterious function. I wondered how it would look if I were to be seasick whilst doing it, and I felt badly in need of the sort of weather-beaten petty officer whose knowledge is yours for mutual respect.

The *Monarch of Bermuda* didn't have a foretopsail, and so there was no need for lots of sailors to be mustered to climb up and furl it; but some Order of the Day appeared about lifeboat stations which put lots of men and myself in the foreward lifeboats. From this and the confusion that reigned in the front of the ship when a practice alarm went, I deduced that I was somehow failing to do some foretop lifeboat thing!

This practice alarm went every day, and I hated it. Partly because I was usually the last to find my own life-jacket, and partly because the other three sections, or watches, of the troops on board were all quiet, settled and waiting, while my lot were still drifting around trying to get organized. And my rather unconvincing shout to the Bridge seemed to sound, once again, as if it hadn't quite been my day.

The position also called for an expert in laying out kit for inspection. Why was it that practically every day my lot got lowest marks and came bottom? I pretend to myself that I just don't care about this sort of thing, but I know how secretly thrilled I would have been to have come first - please - just once. I think I form convenient attitudes of contempt to cover my unenthusiastic response to challenges of this sort.

On the other hand, for some strange reason, I didn't seem to have any trouble getting some of the wilder and more discontented corners of the troop decks to settle down for the night. Consequently - and I was more than a little pleased - I was asked to do twice my share of rounds. It restored my rather battered ego. Of course I realized that it was partly, even mostly, a rather drunken respect for the uniform - but also, perhaps, for authority used in that relaxed way: where a sense of humour is just breaking surface, and where there is empathy somewhere in the tone of voice.

Fortunately, the troops didn't realize that I was terrified of being suddenly seasick whilst doing these rounds. Where could you go to find embarrassment like that? I could imagine the cruel applause - but fortunately that disaster never broke surface.

The voyage took six weeks. We detoured so far that I don't remember thinking much about any wartime hazards. I find this now to be quite a strange piece of soporific unreality, related mainly to boredom, lack of imagination, or perhaps to an introverted preoccupation with disillusion of some sort.

Neither do I remember the arrival at Bombay. But I do recall how quickly it became apparent that the British residents could talk of nothing but the defects of their Indian servants and of Indians generally.

Most of us, including untravelled mechanics and landing-craft crews, reacted instinctively against this barrage of colonialism. I noticed very quickly that my landlord - a great talker - kept his 'boys' waiting and working on and off all day and well into the midnight hours.

The Two Himalayas

When some kind of unrest and rioting occurred in the poorer end of the town, the senior 'boy' - an employee of some fifteen years' standing - was still not allowed to go off about nine o'clock in the evening to see if his younger son was in any danger. This infuriated Jock and me, and we saw fit to tell the landlord what we thought. Jock was my Scottish co-lodger. He had a bit of difficulty with his readily-aroused temper over unfairness, and the 'boy' was rather hastily allowed to go off.

Jock was particularly incensed by the treatment given to the poor half-starved horses that pulled the gharris (a sort of cab), and one day - to my very great pleasure, I must admit - he laid one of the drivers flat out on the pavement. That horse got a rest for a bit; but I suppose might have got beaten specially afterwards.

Their treatment of these pathetic animals was our biggest real criticism of the locals. Jock and I used to go out together and hire a cab for the express purpose of watching the horse stand and have a rest during the paid-for hour. Whether the driver could even see our point was very doubtful. After all, everybody watched a continual stream of suffering. It was hardly even a case of passing by on the other side, because there was almost sure to be suffering of some sort on that other side also. Facilities for helping one another are of course meagre with so much poverty; but care and compassion seemed lacking in the first place.

The worst example I ever saw of this - or of its results - affronted me on my second day there. It is really very objectionable, but there may be some point in at least hinting at the state to which poverty and ignorance and an uncaring philosophy can reduce a city.

I turned round, and saw the back of a big and unusually well-built man walking down the middle of a crowded main street. He was naked, and the medically impossible seemed to have happened: disembowellment - his entrails dragging along the road - yet his limbs functioning. His mind must have gone. I won't go into any more details. I wouldn't expect you to believe me if I did. You'd think that somebody would have done something about it at some stage or other, but apparently not. Everybody just carried on, seemingly regardless. Anyway, half of them seemed to sleep in the streets cheek by jowl with mutilated beggars.

It is better not to try and figure out how there can possibly be a neat dividing line between the filth of these nasty streets and the inside of the attractive-looking hotels. And as you wandered about the markets, aware only vaguely of the anodyne effect of the heat on your stressful antiseptic Western ways of thought, you found in yourself a desire and a thirsty longing for the water-melons and similar fruits cut open to attract you. They tempted me - against what I thought were rather ridiculous official warnings. After all, if you take a melon home and cut it open yourself, how can there be anything wrong with that?

I heard later that these and similar luscious-looking things prospered best in the back yards of villages, where the natives did their morning 'business'; and that a piece of candlewick stuck in them and in the ground functioned as a sort of auxiliary 'root'. It is not exactly Nature's intention that its disease-filtering mechanism should be so neatly by-passed in this way. If it is true that this went on it probably accounts for one, if not more, of the three different kinds of dysentery with which I finished up.

Which kind came from Sierra Leone, which from some melon, and which from the huge rubbish dump just outside my office window, at the docks, I do not know.

The Two Himalayas

I'm afraid I got into a careless, rather dangerous attitude of mind. In my office I used to count the flies crawling up each forearm, almost enjoying the friendly little tickling feeling and just not linking them with the poisonous rubbish tip.

Anyway, there was a lot of boredom. There was no danger of my being able to take a Wren out for an evening. The competition even among the dashing, athletic, 'He-man' types was acute. I gathered that it was common practice to keep a gharri for the entire evening, rather than risk her displeasure by keeping her waiting while you shouted one down. The attractiveness of the posh hotels, luxurious swimming pools, etc., was to me rather hollow.

Somehow or other the odd chance of borrowing a horse for a canter round the local Rotton Row came my way. Feeding stuffs were in short supply, which is partly why somebody gave me the animal to look after. I accepted the gift, saying (and meaning) that it would be great to be able to offer the chance of a ride to chaps off visiting ships. I even thought of getting one or two more horses for this purpose. But this one proved to be enough. Having organized a stable and a syce (the native stablehand-cum-groom) for next to no cost, I then learnt that the horse was a vicious beast which bared its fangs and shot out its head full of protruding teeth in a sudden and most unhorsey way. I think it was irritated by the flies.

Apparently the previous owner had been catapulted off at a corner of Rotton Row, and been killed. The only person who could manage it properly was a French cavalry officer from some elite corps or other. I eventually shared it with him; but it was galling to see him control it with one gentle finger. When I mounted, it took off directly the syce let go. I think it felt and resented the difference between us.

Its name was Panna Prince. Its gran'dam had been the winner of the Triple Crown: the Derby, the St Ledger and the 2,000 Guineas - in the same year. His breeding came out in the fact that you didn't hack - you glided, hardly feeling his feet touch the ground, except in a light, springy way. Unfortunately, he needed resting. Something had gone wrong with his wind; I think he'd been raced before he had got used to the climate, so that he'd had to be tubed.

I sent him out in the care of the faithful syce to some nearby sandy coast. Every time I went out there, he bolted; but he had to stop because the sand ended in a long spit jutting out into the sea; otherwise I should never have dared get on him. I somehow dared that much of myself. It wasn't quite me, but I felt that I ought to.

Nor was it me at all when, one day, returning more quietly from the spit of sand, I found myself cantering into the outskirts of a prayer-meeting. When I recognized Mahatma Gandhi I felt that I must have confirmed the worst beliefs that any of those assembled people could ever have had about the British - but Panna Prince had me away before I could attempt any apology. I only hope that Gandhi, as I rather imagine, chose the incident to reveal or illustrate some profound forgiving thought to those followers around him. Unhappily I never heard.

I hope that I made up for this later, because, when by chance I rented one of a group of modern little bungalows somewhere in the hills, I found that Indira, daughter of Gandhi-sympathiser Pandit Nehru, had the one next door. She was in her twenties, quiet, well-poised and with a depth of feminine aplomb, revealing no turmoil from her earlier imprisonment and political embroilment; as if (presumably wrongly) her involvment was more a consequence of family history than a commitment of her own.

The Two Himalayas

Staying with her - or perhaps next door - were her brother-in-law Huteesingh and her two small sons, Sanjay and Rajiv. The two boys had me scratching my head as to how to play cricket: and somewhere in my loft is a childish present from them.

Huteesingh intrigued me. He had just been released on parole from imprisonment for some kind of passive resistance, which I think was a fairly new Gandhi-inspired conception at the time. It was of course resistance to the British Raj; but, contrary to the image of such rebels prevalent at that time, he seemed to have a most attractive mind, was quiet and pleasant and modest. Not knowing whether I ought to be a bit furtive about it, I walked through the woods with him, trying to keep an open mind and not to offend him; and not finding anything to take offence at anyway.

When he told me that I was the first Englishman he had ever been able to talk with properly, I hoped it was an exaggeration.

I don't think I was all that gullible, nor was I anything like a left-wing intellectual; but I did wonder what sort of stupid, prejudiced bastards he had come up against.

Some long time after this - when sitting alone in a big native tea-shop-cum-restaurant at the Kashmir border, that same young lady, Indira Nehru, recognizing me, came over and asked me if she could join me. She was, surprizingly, travelling alone. I felt honoured in a quiet sort of way, and was again most taken by her demure behaviour and complete lack of any pretence. After all, she was Pandit Nehru's daughter, although possibly unrecognized by the passers-by.

In contrast, an 'informal' tea I was invited to on the Governor of Bombay's lawn about the same time was nothing like as relaxed. The 'unspoken' in the teashop was kindness and human acceptance; on the lawn it was kindness and rank.

No-one would have predicted that this unpretending, unobtrusive young lady was in fact to be called the 'most admired person of the year' by a Gallup Poll in the USA in 1971 - and was, of course, Prime Minister of India for so many restless years of its history. At the time, she gave me the same feeling of quietness as those naval destroyers that I remembered lying alongside at Londonderry.

Her later history, and the fate of her family in more recent times, seems a far cry from that tranquil, passing, Kashmir scene.

But I am jumping ahead. My job at the little one-eyed landing craft basin needed somebody with a bad temper. I had flared up once or twice in Londonderry with good results; but that was to get ships to sea. Here in Bombay I had no such incentive. My natural conscientiousness thought out unassertive ways of plodding on. So, while I suppose the bugs were ruminating inside me, I prepared painstaking inventories and other rather pathetic preparations for the future smooth-running of some nebulous Indian organization that never got off that intractably White Man's ground. Some of the paperwork was torn up. Some didn't even exist, like that required to dispose of an unwanted fleet of landing-craft. Clouds of dissatisfaction hung about in London, Delhi and Bombay, which each had a different policy. Instead of slamming a few doors, I worried.

I needn't have! With confusing irony, with majestic stupidity, Calcutta eventually sent their landing craft by train to Bombay - bilges full of mosquitoes - to be dropped in the water and sunk. Perhaps the water was better there? Perhaps our paperwork was better, a thousand miles away? Perhaps that's why it got torn up?

The Two Himalayas

My mind wandered continually away inland - introspectively inland - away from these thankless preparations and from the superficiality of the bars and the repetitious all-male social life. The usual thing was to go off to a hill-station for time off, but these made me miserable, drinking on the boring fringe of such life as there was; indifferent health tips one towards a sour-grapes attitude to the more boisterous.

I used to go off with Bill and his wife, local residents, on shooting expeditions. I never quite matched up to it all, but he was very warm-hearted and tolerant of my efforts. It was typical that a deer jumped right over my head, which was tucked under the hedge while I was thumbing through his little guide: *Shooting and Shikar*. The Indians thought this a huge joke. "Sahib in book - not see deer - deer over Sahib's head!"

I could just about get up the hills, but the Indians ran continuously, far ahead in wide sweeps, beating the jungle towards us. Surprisingly, they all carried bows and arrows. They also carried chapattis stuffed into their filthy loin-cloths. I discovered this storage system one day just after I'd hungrily eaten one. The memory disturbs me still.

Once or twice I wandered off, for days alone, borrowing Bill's rifle, with which I shot one pea-hen; and with which I once found myself caught up as the 'White Man' in a jungle story that might have come from the readings of my youth.

As I approached a village in the half-dark, patches of the jungle had been lit up, every branch outlined clearly in fireflies, glowing themselves into a memory, a mellowing delicate tracery. I sat amongst the villagers drinking tea, wondering why this part of the forest was dark; but still bemused by that strange glow.

Yes, there was an elderly Sahib who would put me up somewhere nearby.... But two nights ago a woman had strayed just outside the village and been attacked by a tiger. They all seemed very excited. It didn't happen often. This tiger - a man-eater - would soon be getting restless again. And they poked the fire to keep the dark shadows of the surrounding forest away.

I was in an impressionable mood. If they tied up a goat tonight, would I try and shoot the tiger when it came back tomorrow night? I wasn't sure about the logic of their plan but they offered to build a little platform up the tree for me, which I could understand all too clearly. Indoctrinated, I presume, with past years of the *Boys' Own Paper*, I said that I would.

So it would be necessary to buy a goat: and I would pay the 35 rupees? Yes, I said, no longer very interested in my financial position, which I saw as perhaps taking on a hypothetical turn, with its dubious relevance in one's last few hours in this world.

They took me to find the Sahib's house. I relayed the story over my evening food. But the very doddery old couple retired early, leaving me to spend a gruesome night in a half-lit library and a rackety old bedroom.

Eighteen feet! I came upon it at last in a musty old volume - the height that tigers can jump. The jungle around wasn't all *that* high, surely it wasn't? I cursed the shadows and my imagination and my poor memory for the heights of trees. I had forgotten the fireflies, and only remembered the dark.

Towards the end of the next day I realized that the whole thing had been a con, to get that 35 rupees out of me. I don't remember the villagers' explanation but I doubt if the goat ever existed. I just remember my relief!

The Two Himalayas

One weekend I went to stay with one of Bill's acquaintances, a rather elderly lady, who lived by herself in a little village that occupies the gap in the Ghat hills some eighty miles behind Bombay. The city's only main railway line, its water supply pipes and the high voltage electricty pylons all squeezed through this gap, hugging the steep side of a gorge in the hills. Her house was full of old hunting trophies, tusks, and so on, hung all over the old timber walls. She had been a big-game hunter in her own right.

Just after I arrived, the District Commissioner telephoned her to say that two parachutes had been reported dropping somewhere near this gorge. As she was the only one in the village who could speak the dialect of the tribe that lived down in this thickly-wooded, remote valley, and fearing an attack on the lines, he had little option but to ask her to try and find out if the tribe had seen anything. What else could he do? It was a matter of gaining their confidence, which he could only really entrust to her.

She decided to climb down into the valley at daybreak, frail as she was, but trusting in her years of living in forest and jungle. Here it was again, the same kind of tiger-dilemma. How could I let her do this alone? What use I might prove was another matter - but she sorted me out a gun, and I decided to overstay my leave.

How romantic this would have been had she been wild and young and beautiful. But, although she was far from that, the idea appealed greatly - the old-fashioned chivalry of the story books. I glossed over the practicalities of a heavily-armed Jap and myself encountering one another, and couldn't wait for the morning.

Once again, however, it petered out as a false alarm. I went back to Bombay with the boyhood's-dream part of me secretly frustrated; but I was, I suppose, alive.

Panna Prince gave me opportunites of breaking out of the town and away from the futilities of trying to graft an RIN organization on to a (sometimes rather racist and 'anti') RN one; away from that thankless atmosphere, where all the uncertainties of war were continually changing the ways in which London's view and Bombay's and Delhi's didn't fit.

I was invited to try him out with the Bombay Hunt. Surprisingly, this was a very free and easy, casually-dressed little affair - completely different from the painfully first-class Yacht Club. They hunted jackals in the hill country some thirty miles or so out of town.

Poor Panna Prince started causing concern on the first morning, tripping over bumps and stones directly he left the station yard. But after about an hour he seemed suddenly to realize that the whole world was not turf, and after that he hardly ever faltered again. I thought this sudden adjustment, after a lifetime spent on very special smooth surfaces, really quite impressive. He quickly learnt that cactus was nasty stuff and that prickly hedges of it separated fields at different levels. He didn't screw up his face, as I did, just jumped and landed as if born to it.

I'm afraid that I couldn't stand the pitiful sight of a poor, half-exhausted jackal, although they are certainly not very endearing creatures. Having once succeeded in putting the hunt off the track, I went out with the secret intention of rescuing some other hard-pressed beast. Hypocrisy? A sort of Judas? Perhaps; but as everybody always went home immensely pleased with themselves, kill or no kill, why then shouldn't he escape?

The Two Himalayas

The particular one that had first clinched my sympathies had stood gasping no more than six inches from a hound, both just against a boulder, rounding it from different sides, exhausted, nose to nose, each unable to do anything but pant and stare. I just happened to be some ten yards off. I just stared too, happy to see the hunt veer away. On what possible grounds should a chorus of death come to surround such a pathetic, beleaguered little thing?

A camera could have preserved that rare little gasping confrontation. Or is it rare? I only ever read *Country Life* in the dentist's waiting room, but maybe I should join their staff - because I once saw a fox jump off a railway embankment in Surrey on to the *backs* of a densely-packed collection of bewildered hounds, and from one back to another the resourceful little devil scrambled off to escape behind a hedge. I was just out cycling - not part of the hunting scene. Perhaps I'm just a hunter's jinx.

One or twice I sent the syce into the country with the horse, and camped near a village. We would build a very simple compound for Panna Prince, while I slept in a hammock. I remember how, when I woke up, the natives who had slept in even more primitive conditions in a neighbouring field were shivering violently and clearing their throats in chorus.

They cooked one of their precious chickens for me - or started to. I ate it, I had to, raw and all 'gnorple', if you know what I mean. And there was orange drink 'flavoured' with the cheapest of those revolting scents favoured by girls who sometimes sicken the passer-by on a London pavement.

Once Bill shot a panther; it was near a clearing where they were brewing jungle gin - whatever that is. Bill over-indulged and passed out, leaving me with the dead panther, a knife and two typed foolscap sheets of instructions for skinning it. I must have been pretty tight - I don't remember any more. Some difficulty in concentrating mercifully blotted out the details.

He had previously shot a wild pig that day, and somehow I ended up with the head sticking half out of a sack. I cannot remember whether it was the head only of a medium-sized pig, or whether I had got the whole of a poor little one; whatever the case, what followed appears to me now to be so thoroughly objectionable that only gin can explain it away (I don't mean excuse it). Later that day, apparently lacking all sensitivity, I decided to board a tram, full of Muslims, that emptied itself, on religious grounds, out of the front door as I sat down with the pig at the back. The dilemma as to what I should do then was very real, but my religious sensitivities went under with my amusement at their expressions as they shuffled off. The fringe of imagination might just be able to grasp that sort of disturbance of the feelings - it was funny, a naturally funny situation on a religious tight-rope, and I hope I tell it understandingly and inoffensively. Words can't quite do justice to that tram-load of retreating forms.

There seems to be more from this time to be ashamed of than just a dim sense of religious decorum. Bill was different from me in many ways. He had a way of getting me to do things against my better judgement; so I never managed to get him to turn back on the Friday night when the whole Bombay Docks area, miles away, was obviously ablaze, lighting up the horizon with a huge fire. Yes, he was different from me. He couldn't have his plans spoilt, and so we drove on into the country.

The Two Himalayas

On the Monday I got the criticism I deserved, and felt it badly - because I had longed to turn round and rush back and would normally depise a Base Officer who hadn't. This wasn't exactly what my reputation wanted - and a slightly nastier type of worry set in. An exploding ammunition ship had blown two smaller steamers on to the quayside. As I watched the debris and floating bodies the next day, I suspected sadly that somebody else had perhaps filled in on the electrics for me . . . .

So, when the RIN staged a sort of half-cock mutiny, I hoped to be able to make up for it all somehow. But there was no chance. As if to make me both ashamed and proud at the same time, a Royal Naval cruiser steamed very, very slowly into Bombay harbour, no guns pointing - just the ship's company dressing ship; lining the decks at attention as she came in. After some days of tension and uncertainty, the effect was quite memorable. I think many of the Indians found themselves smiling; this tangible, infectious thing, *discipline*, sensitively used, can do to men what words or excitement cannot do.

I know that it produced a lump in my throat. But I can't really figure all this out. Quite inconsistently, I hate being lined up myself; but secretly, I would have joined that crew. Perhaps, also, I was homesick for a youthful naval dream - knowing that it had all gone wrong somewhere.

Which reminds me that the cruiser I could well have been destined for was sunk out East, later on in the war. I must have been fed-up to have applied for it, to have contemplated trying to figure out electrical problems while at sea, perhaps sea-sick, wet and cold. Some illogical streak in my make-up finds electricity an absolute incongruity on a ship at sea - the antithesis of all things nautical, and as repulsively out of place as seamanship, rigging and lifeboats are in place.

I realize that anybody who has been to sea must know that I need a good drenching! but I *had* been ejected from the wrong job I had loved so much at Londonderry, and the punishment had started to bite.

4

It all happened fairly quickly. I was called to New Delhi for informal discussions concerning our sections's needs and duties. In the course of our chat it soon became apparent that Watson, my side-kick and very much my old friend and trusted confident, who had worked with me for so long on the base, was going to let me down in order to get promotion. He had been a real, warm friend, but there he was condemning all that we had discussed and shared together. I felt pretty sick, and saw that I was alone in my loyalty to the Indians I had been sent out to train.

My carefully thought-out report came back. The Captain of the base had scribbled across it in red ink. Apparently this carried nasty implications for my future, if any - a very rarely used form of censure. "The worst form of journalese" he had written. Perhaps it was.

I was quite down now, and must have been rather fed up.

And with ironical timing, I was due to ride in a mile and a half steeplechase at Bombay racecourse. This sort of thing was really well beyond my capabilities, but someone had dug out Panna Prince's pedigree and, not knowing my riding and quite unknown to me, had him down as favourite. There it was, in print, in the little books issued by the tote.

One dismal early morning practice convinced me that somebody else had better seek that way of achieving glory - although, God knows, I could have done with a boost of that kind!

Instead of the glory and the boost, however, I got a bit of kindness from a good-natured Commander somewhere above me, who reckoned I'd better see the MO. He was right; jaundice and probably dysentery.

Somebody else rode the race. Panna Prince went lame. I went to hospital. And so began that unremembered sequence of hospitals and leave camps that seemed to emaciate my mind as illness gradually emaciated my body.

It was all a very long-winded affair. They started by chiselling away at the wrong bit of my nose - if indeed there was a right bit to chisel; leaving a piece of cotton-wool dangling ominously.

Apart from that, they really didn't seem to know what was wrong with me, and so didn't know what to do. I became very fed up lying around in hospitals.

Somewhere or other I got friendly with the pathology staff. There was a long expensive-looking room equipped with sophisticated microscopes, and so on. One day, while I was twiddling the knobs, examining some micro-organism or other, a nasty-looking, shaggy old goat wandered in at one end, found nothing to eat, and left

The Two Himalayas

through the far door. Nobody took any notice. I stood guard over the little glass slides awaiting tests. One of them was mine. India, nineteen-forty-something!

After a time, when boredom was becoming an acute problem, the faithful old syce craftily set up some early-morning rides for me. I'd hop out of bed about 5 a.m., creep through a gap in the hospital hedge, catch a tram to and from 'Rotten Row' for my bit of a canter, and creep exhausted back to bed.

The nurse had to rub my behind every morning to keep away bed-sores. My inexpertize at riding must have shown, as she smoothed out the bed, muttering about the strange imprints on my bum. She looked bored too - I suspect that life off duty was very much more interesting than my behind. I was her only patient, and her boredom made me sad. She was, I suppose, also rubbing in the derelict sadness of my position.

At some stage my delightful old Sikh havilder [sergeant] took over the cure for the jaundice. He came across the city each evening with something rather like yoghurt that he called *Chah'h* and smuggled all the day's hospital food out. I got cured of the jaundice in a matter of days, leaving everyone else still yellow and languishing; but the dysentery remained.

I lost the best part of a year lying around, or as some sort of walking patient, or convalescent. Perhaps I went back to work for a time; the heart had gone out of me, and I don't remember.

I do remember that Panna Prince became a problem, which I was stuck with while lying around in some Nissen huts 'under investigation'. I remember that ward. I stuffed cotton-wool and vaseline into my ears because I couldn't sleep, and still I tossed and turned. I nodded off after two or three sleepless nights and was woken up to have a sleeping tablet. I was psychoanalysed and sigmoidoscoped - both by charming young lady doctors (a stunning innovation at the time), whose presence was our one and only delight.

The day that I saw the psychiatrist I was feeling unusually perked up because the Nizam of Hyderabad had met me that morning and offered some fantastic sum for my horse - I expect the psychiatrist gave me a clear eight out of ten! A posh jeweller had answered my advert and arranged the trial ride.

The Nizam was a young man, pleasant and fit and seemingly Europeanized - but nothing happened. I heard no more. I was going to write that Nizams can be utterly infuriating, but gather there is only one, as the title is reserved for the immensely wealthy ruler of this one state of Hyderabad. He, or his father, was reputed to have been the richest man in the world. The Indian Government had banned the sale of the family jewels: the Supreme Court of India could have been the only possible auctioneer.

I was to learn that these rulers have - or had - a special kind of pride, which includes finding it *infra dig* to say that they don't really want a thing. They must have a convoluted sort of ego, which I suspect makes them pretend they don't soil their minds with money's-worth considerations. So they just let you down.

That, at any rate, is what I was told, and right or wrong it made me hopping mad. I decided that I wasn't going to be treated like that, that he'd got something to learn. So I went to his Bombay palace (in civilian clothes of course, to keep any funny Naval angle out of it).

It was a vast building on the hill overlooking the town. I asked some flunky or other if I could see him. On either side of a great big marble entrance hall were two

The Two Himalayas

elaborately dressed guards or retainers, big fellows. They refused me entrance, saying that I wasn't allowed through the harem on the other side of which was, presumably, the Nizam. Rather stupidly, I accepted this unlikely domestic arrangement, and gave up without causing any ripples, afraid of I don't know what.

I rather think that this flabby piece of behaviour on my part incensed me more than ever. The last I saw of him was leaving Bombay Station on a private train. I thought there, at least, I might get at him, somehow or other, but I couldn't find a way through all the mystery of the private train set-up. I suppose that, both metaphorically and literally, I flopped back into bed.

I suppose too that it's in bad taste to record why (together with one or two others in the same boat) I had to rush to the cold tap over stories we heard about the young female doctor's extremely necessary professionalism in this sigmoidoscopy business. I am sure that it was all strictly in order, but she did expect you completely to forget your modesty. And rumour had it, lady doctors being so new and frustration being so widespread, that you'd get put on a charge if you lost the struggle of mind over matter. Oh Lor', and she was so nice, and I'm sure she didn't realize how soft and rare and tender her touch was. Oh Lor', perhaps she didn't realize either how youthful desire can surface even in illness.

The poor horse and my finances still worried me. The only British (or near British) person interested was a fellow who ran a private army for a prince somewhere up Kashmir way. I wouldn't sell locally, because I feared the gharri-type ill-treatment for the horse; worse even than butchering.

The old syce was sad but agreed to accompany Panna Prince on the thousand mile journey on the train. The cost of the horse-box and fares, etc., was astonishingly small, but nobody knew how to organize the transfer from road to rail. Eventually they agreed to stop the Rawalpindi Express at a sort of miniature Clapham Junction in the Bombay suburbs. We got the horse in the box in a siding, a great crowd watching. The express pulled up and waited... and waited... But such a departure from routine was too much: I had to work the points myself. The syce burst into dignified tears, and the train pulled away. Somebody there must have known how to couple up trucks. That was the last I saw of my horse. My conscience was not quite clear about it. I only hoped the flies were less cruel up Kashmir way.

Some time later I was sent up that way myself to recuperate. The last part of the journey was in a lorry full of noisy troops. At one point their perpetual back-chat and cross-talk suddenly stopped. I followed their gaze and looked out - it was a stream with the green look of England about its banks. That sudden special quiet, that long unbroken quiet, said it all.

Kashmir itself was refreshing, green, beautiful, with wild flowers and again that lovely thing, a stream with weeds and little beetles to stare at. Although some of the natives were rough and aggressive - physically more like heavily-built Tibetans than Indians - the place fascinated me, especially Srinagar and its houseboats.

I got to know a local man who ran a herb farm. He wandered about the hills collecting wild herbs which he sold to the herb-minded folk of India. That sounded idyllic, a quite perfect life, and we talked of my joining up with him after the war. Such is life! It suddenly came to light that his partner, who had recently died, had very

The Two Himalayas

probably been murdered by the natives - something about powdered glass in his food! I dropped the idea.

I might indeed have thought of climbing hills sometime in the future, but the fact is that I was really very weak and any such thinking was tinged with unpleasant suspicions that I wouldn't be around anyway. I suspected that other people thought that I was probably a gonner, but I don't think I ever fully faced up to that. I just remember a special sort of staring into the stream, my mind blank but concentrated.

I remember staring the same way at the ceiling some time earlier, on the night when VE Day was declared, and I was more or less alone - 'strict bed' - listening in that same way to the noise of revelry. The young lady doctor, who had that very morning forbidden me even to cross the ward to the loo, came tripping up the stairs and dragged me down them, and round the hall to the dance-music. My legs went all funny, but there was no more staring. She was a little tight; but it put me right on top for a while. She honoured me with a little story of a female patient and the suppository that had shot out when she saw a spider!

Since then, I had stuck at 7½ stone - and being six foot tall makes that rather skinny. Although I got to walking round a bit, I had little enough to encourage optimism. I can't remember much about it except that I tried water-colours, and sending things home. I expect I tried thinking, and as near as I could get to the opposite of thinking. I sometimes wondered if I would ever see Molly again, but I was muddled about whether that would be sensible, or even kind.

I ended up in an RAF leave camp somewhere in the mountains. They were mostly too boisterous for me, who hadn't much boist left. The only positive thing I really remember was trying to prevent the WAAF girls being swindled by the pedlars. I could manage quite a little Hindustani by then, and enjoyed that as the nearest thing to feeling a bit of life, and involvement in life.

I would occasionally mooch off, frowning and chafing at the net of failure and poor health. There were no decisions to be made - hospitals protect you from the worries of the outside world. A demanding, hairy situation would have been a tonic. I became more than ever preoccupied with other people's opinion of me, knowing that anything colourful that had ever happened to me was really only because of my youthful ineptitude at ball-games. This had forced my life into a non-standard pattern, and had formed some sort of image in people's minds. But now it looked as if my thin body would scarcely be able to cope even with routine - an identity-less, unnoticed, untalked-to existence. And so it would go on . . . if it ever did . . .

Underneath I was in a perpetual turmoil; longing for a decent reputation: to be a somebody not a nobody. If a 'somebody' took some sort of notice of me, then all was fine - *if* they did. If I was 'left out of it' then I was glum, however much I pretended otherwise. These secret preoccupations were on the go most of the day. I knew that girls scarcely noticed me. We had to be thrown together by circumstance, continually in the same room almost, before I could get anything like a smile out of them.

On that basis I knew no hope of happiness other than that somehow I might blossom out - not just into a nice person, but into an attractive one. One that attracted, presumably, a soul-mate.

The Two Himalayas

Uniforms attracted - but I'd failed even in uniform. I'd put the uniform, as it were, on that secret streak of self-reliance that I used to possess, and it had all gone wrong. Religion? One connected it perhaps as a last resort for a wash-out like me; but I'd gone into all that rather exhaustively over the years.

It seemed that earlier rejection of me as a youth was returning in some ominously finalized form. It had dragged out over saddening months, and was keeping me away from even the eddies of life, while inside I was bursting with the wish for involvement.

Quite recently, some thirty-five years later, I was stared at by a complete stranger, who thought she recognized me as the tiresome skinny wash-out in borrowed RAF uniform who had to be put to bed secretly one night in their school ("can't bloody well get back tonight", and "can't you get it some bloody straw?" and "shouldn't have borrowed the bloody pony . . ."). Except that he had looked a bit of a gonner - a bit of an unlikely relic from the Navy, invited by mistake.

I didn't know what she was talking about, but there was my name in her old scrapbook. And there was a photo of herself, a stunner, lovely, miles from anyone; yet unremembered even by lonely me: so twisted was I out of every hope - oblivious to female beauty, swearing in unfamiliar mixed company.

I don't think seven long months weighing 7½ stone lends itself to much else; and even when I did register 7st 9lb, and then a week later two pounds more, I still fretted along very much the same lines.

I didn't give much credit at the time to the new RAF doctor who (I have since come to think) initiated this improvement. It was the sort of 'cure' that the medical profession in general discounts, like the *chah' h* that cured the earlier jaundice - or did it?

He gave me two little pinches of some tasteless white powder, which I found quite comical; it was homeopathy. It wouldn't attack the disease, he said (an increasing assortment of dysenteries as it turned out later), but it would assist the body's natural curing reaction to it. You tell me. I know that I was nasty enough to try and attract attention with cynicism about it later that evening in the bar, and then forget it, reconciled to misery. Yet I gained weight, more and more.

I wrote earlier something about religion as a last resort. There is indeed a tendency to veer that way, hoping for some thought that will have some real lift in it. But I'm afraid that the more appropriate angles of religion seemed rather empty and hanging in space in an almost contrived sort of way. Ideas of eternal life, resurrection, or the more Hindu reincarnation don't really capture my interest. I don't say that I find them impossible, it's just that they could only hold my interest under stress.

'Eternal life' to me is quite divorced from the clock, and more a mental refuge for the fact that spiritual qualities can't be bounded by time or by our powers of understanding. But I have no difficulty at all in believing that Christ was the best interpreter of these things and was utterly convinced himself.

At that time it was all a very soggy mental area and certainly not the place to find peace, having very little effect upon waves of restless thinking. It was more (far more) to the point that I saw no prospect of impressing the world around me enough to capture sufficient of its attention, and one of its females, for my happiness.

So I mooched off now and again, on an old pony, to the nearby village, where the locals were friendly and would talk a lot about guns. Little did I realize at the time that

The Two Himalayas

the innocent twinkle in their eyes reflected their plotting minds about the coming Indo-Pakistan troubles. Rather it flattered my ego that they thought I might, perhaps, get a gun or a licence for them. Even that little boost, and their cups of tea, were welcome.

I walked there one day, but wanting, perversely, just my own company. Nobody in the camp understood me. All was so superficial; perhaps I was, after all, my own only friend? It felt like that. I was a fed-up skeleton. Mentally, even the treadmill had stopped - and I had fallen off.

I drifted aimlessly up the slope at the end of the village, away from people. I bought an old-fashioned, faded little book from the last market stall. There was a gate ... so I sat there, as people do sit on gates, thumbing through it. Tibet, in the distance over those foothills, had once not long since echoed a sad and questing me. But it looked down now, emptily unaware - almost mocking. A lump came in my throat. I shrugged my shoulders, as if to spill the dregs of a castaway me ... a failure, mistakenly uniformed, dumped back in a sea of failure. I sat there silent, yet somehow aching.

Then, hardly aware of it, I felt a stirring - as if a wavelet had run across a pond and moved the silent reeds. Some gentle wind, like a little wave of compassion. A stirring from something that had strangely taken root in my boyhood? Long-forgotten words floating from the book?

I had once been a choir-boy, secretly loving those words - 'Ye shall find rest unto your souls' ... they were an echo from the memories of far-off, long ago. So I sat there dreaming ...

That was it, wasn't it? That was it ... 'learn of me ... Take my yoke upon you, and learn of me ... and ye shall find rest unto your souls ...' And slowly my eyes came back from that meaningless Indian greyness; came back from that still-remembered muddy track that wasn't their England nor the way to England, and gazed back through that white meaninglessness of a book's pages to find that random sentence, the old book disconsolate on my knee. And as Time came back, I suppose something began to matter and to feel its way into me. I don't know how something stirred, nor why.

Rest, it focused, was something to be *learned*. Could that be?

Because it was so odd, I somehow knew that here was *truth*, a sort of improbable Galilean truth about that Galilean calm, that wouldn't run away, that was unchangeably learnable ... 'Rest', anchored strangely to schoolboy familiarities? To learning?

I felt something tapping at the real Me, saying that here were more than thoughts, here was home; a new life to awaken to, in a past-forgetting, reputation-forgetting land.

Perhaps a part of me *had* sunk? - and cleared a kind of log-jam?

My eyes ventured on to other words, and they stared kindly where that deadening fed-upness had been; and they survived and soaked in. And that is how peace found a foothold. It is how that faded book befriended me.

And those ever-eddying thoughts about my doubtful uncertain reputation had sunk among the shallows, out of sight.

The idea came later that restlessness could, after all, be *caused* ... but I am uncertain where that fits in; and it was only later - strangely, perhaps - that I found out that the 'yoke' ('Take my yoke upon you') meant a way of carrying a burden - not the hated Roman oppression of the Jews.

Perhaps it is a little simplistic to say that one sentence, one thought, 'Take things as I take them', was far the biggest watershed of my life. It seemed to me that, at just this point, I was scraping near the rocks of reality of something very fundamental.

Take things as I take them . . . and you will find rest?

I would like to have laid this out for you, this not quite old-fashioned chapter of another book, as a beautiful thing from a wonderful mind. I hope nobody will think that I became too overburdened with the 'religious' thing: because religion (or not quite religion) is mostly hull-down over the horizon, keeping its head low for most of this story.

I wasn't 'saved'; but my brain was suddenly brought to an intense rest. Wherever thoughts come from I do not know, but from then on they started out differently and took a different path. I had fallen in love with being 'of no reputation'. I sat there soaking up odd bits of this hope of peace - the hope of spiritual peace.

It was something, I gathered, to do with an attitude, but was tied up somehow with a law. Elusive, but there all the same, it drew me because I was hungry for it; and life went over into a mellower key. I had escaped from the way I had been taking life - which was causing the spiritual soreness that was *me*.

I don't know how long I sat there, nor how many days went by - but I look back now and see that I had found a window on to a deeper harmony - discovered by some uniquely probing mind.

So that the spiritual no longer clashes with the natural; and above all, perhaps, could there not be some kind of inner rest for those who sit there, unconcerned about 'keeping their end up'? Could there not be some law that Jesus instinctively anchored Himself to?

' . . . take things as I take them, and you will find rest.'

He made himself of no reputation - and showed us 'rest unto our souls'

. . . and somewhere in all this is 'cause and effect' - some kind of rest or restful reassurance for those who latch on to the background instincts of Christ.

I could not then have conceived that this was to grow and become interwoven with that earlier type of fulfilment of which I used to dream - me and a girl on a little boat, clothes folded away in a baptism of trust; sharing sunshine, sails, ropes and tiller; because nudity, young unsophisticated nudity, was part of my dream world: a world of quiet anchorages, somewhere awaiting the tide; longingly wanting tumescence but, poignantly, balanced on trust. Keeping back the tide within me, because deep down that's the way I wanted it - until the tide had turned a few times.

But that's another part of the story.

I was to stare up many times in those days at those same hills - an aimless figure, wandering about.

'Rest and Peace are but calms in man's inward nature . . .'

'Ye shall find rest unto your souls.'

The phrases tangled with each other. I didn't know which was from the book or which was some echo from the days of my childhood.

The Two Himalayas

Perhaps I recaptured something from each: and perhaps the distant hills said, timelessly, that He wasn't far away, this God. Wasn't far away from our needs, our true and real needs. That once-discarded old book (second-hand, for half a rupee), and that older Book mixed up with 'long ago' . . . Yet somehow not too far away?

So I ruminated, through a few slow days, not knowing that they *were* slow. Unaware of time. Resting my mind in a few slow pages . . . but especially in a growing background realization that behind all this there was a simple principle, rather like 'cause and effect' emerging, as it were, to make our disjointed little spiritual worlds into reconcilable worlds. Worlds as predictable (almost) as nature is predictable. I expect I got it from something in that same book, strangely enough, because it was all new to me.

It was an escape-discovery for me - somehow wrapped in spiritual emotion - because all this seemed to plunge right back through into the spiritual muddle that civilization and the Church had left me with over the years. Years of perhaps cynicism and a growing boredom - sliding into a bit of a pain.

I don't know how much my wanderings and the restlessness of those days was due to my uninspiringly poor health, but I twisted often through that gap in that un-English hedge - through that gap; perhaps into the past, perhaps into the future. But whether my mind was on the past or the future, it seemed to be slowly, inevitably, germinating the same truth, which I could finally see: that the 'spiritual' world was not all wrapped in chance and unpredictability but was just as law-abiding as Nature herself; cause and effect being there, just as in Nature, and even applying to our search for 'rest'.

Perhaps I should have researched that old book more in the first place. I was to find, later, that I was only in fact realizing what Henry Drummond was to give to the young, wondering post-Darwinian world of his time in *The Greatest Thing in The World*.

And I should like to say, even if this *is* all a bit too 'religious' that it somehow gets near the nub of our spiritual aches and pains.

And I find it strange, and interesting, that I - an unhappy wanderer of a later time - should, by an unlikely chance, owe so much to the thoughts of a man who had captured the imagination of an earlier troubled generation. For it was, in fact, Drummond's other book, *Natural Law in the Spiritual World* that first caused a widely acknowledged stirring of hope. Hope of escape from the Genesis-Darwin spiritual enigma of the time.

As far as I was concerned, the Church of my youth may have hauled it all aboard: but seems to have lost it - or nearly lost it - overboard again. And it leaves me wondering whether we could not perhaps have that 'faded old book' hanging figuratively at least on our wall, giving a little spirituality, a little reassurance to the modern fading of our inhibitions. (You will see why I talk about inhibitions later).

I can't help feeling that this all rolls up the clouds in a part of our horizon with an offer of a mental reconciliation between Nature and religion: and, if you'll read on, perhaps a different melting of our inhibitions. Two kinds of peace?

My recurring instinct is to make it all simpler. I suppose I am too acutely aware that all this was while alone in that almost desert, and when probably in a very special receptive mood.

I have in fact spent hours, days, weeks trying to make these thoughts simpler, less 'religious' - more attractive. But I've decided that this is impossible: all I know is that it did things to me. And as I understand that a further substantial reprint of Drummond has only fairly recently been undertaken, it must presumably have done something for somebody else not all that long ago?

One thing I remember is that he apparently came away from listening to a sermon on 'rest'; but still wondering *how* one is supposed to be able to find it. He was thinking, of course, of our restless hearts and minds.

He felt that Christianity overflows with wonderful words, but if one were to penetrate most of our lives little would be found actually to guide us towards these blessings - how in fact to achieve them . . . how to achieve: and how to bring them back if for a minute we feel we have achieved.

He was referring, of course, to a more religious-minded age; but we may be getting nearer the reason why so many of our churches are being left empty.

And this not for the select few with the appropriate temperament - but for all and any of us. And not by some trick or clever knack rumbled by the few.

And so, perhaps, you may have been able to put up with the odd paragraph . . . ?

I don't want to drum Drummond into you, but finding his old book in the setting of those Himalayan slopes is central to my story, and to Ulli's; although I don't think Ulli ever read it. I doubt if I would ever have shown it to him - with its old-style and dated cover and rather Victorian air of all that is outmoded. But one day in the kitchen he suddenly caught on somehow to what I "thought about this religious thing". He said that "it figured". I remember (because it might have been the only time we ever caught ourselves talking about it) finding ourselves on the same wavelength.

Nearer home - in Ireland.

The meek inherit the earth . . . 'They do not buy it; they do not conquer it; but they inherit it,' Drummond had written.

Somewhere amongst my slowing down to sponge up all these thoughts - strangely memorable, a slow, powerful thing - this phrase tangled itself round my conscience; because there was also, as the days went by, quite a bit of dreaming. It is not like me anyway not to have the occasional dream. So 'they inherit it' captured my mind. In a way it shamed me. It seemed to point toward the elusive in a slowly fading Molly, who might now be nursing a special, deeper sadness. It would come to my mind when I thought of her; and maybe because of that she kept coming back to my mind.

I wondered rather sadly about my unhappy contribution to that young life. Some time during the following weeks I began to see these thoughts of Drummond as something that could have joined us closest at the very centre. She was a natural inheritor of the earth. But I? I could not put my past gropings, nor the dawning enchantments of this new time, into words. The change from cacophony into that quieter undisruption had come the long hard way. But I got stuck with a bit of nostalgia, a bit of 'might-have-been' - that we two could somehow have got our finger on the same pulse, yet didn't because of me.

If it seems out of place to bring Molly into these perhaps over-complicated pages, she seemed to me then to be the only person who would have understood me. She came up, too, as an unspoken accusation of my shallowness, as well as my restlessness in having left her. Perhaps it was the hidden irony that wouldn't leave me alone. I had left her inheriting - not the earth - but a bitterness; with struggles that could creep round that virginal soul and wreck her youth, until perhaps she found a sadder peace. I had left her; *pax vobiscum.*

Maybe it would have faded quicker if Drummond's thoughts have blown over; but they didn't. They helped, if anything, to keep alive that rather vexing old idea that it should have been 'Martyred Glen' - that there was some strange meaning for me in the name of her village home. I had wondered, too, whether that glen could be 'unmartyred' again. A nice thought, but hypothetical, Drummond or no Drummond. I had only to look down at my skinny frame and legs. I could be allowed 'rest', but not really hope of that sort.

I wasn't quite sure about keeping up those old memories; but I do know that this element of turbulence, caused by the hopeless practical outlook for my life, was to wither away like the first abscess that Fleming saw wither away following his discovery of penicillin. My spiritual abscess was apparently healing from within. The poison that had created so much disruption was vanishing. And, like Fleming, I wanted only more of the cure. He knew that it was not an opiate. I think I did too. I was just a bit sad about Martha's Glen . . .

Perhaps with some rather devious bit of morality I bottled everything up with the uncommon truism that if restlessness could be caused then so could rest - which was what Henry Drummond was on about; and that was important. I hadn't been quite able to fit it in, but it was not a bad excuse for seeing Molly again one day. A hopeless indulgence, only half genuine if I was honest, to get this rest thing, without the poison, back to the girl it had left.

Instinct, and two or three of her letters, had left me with the picture (now rather ageing and remote) of night-time candles burning late in that tiny room; of someone who had sobbed themselves to sleep with hopeless problems of my making.

And an odd challenge defied me - wouldn't quite go away: should I, *must* I even, show her my skinny frame to cloud her memories with unattractive doubts - to put her off me and prove to myself that it was a *me* now at rest from myself?

But that is reminiscing.

I am not very clear about the next few weeks. Presumably I passed them exploring the uncharted freedom of the quite new mental country I had broken into, where frontiers disappeared and a new harmony emerged. I only wanted the fascination of thinking of it whenever I could. I cannot remember if I went back to work for a time; it is irrelevant, because I would secretly have enjoyed relating it to working life. Even idle thoughts would lock into what seemed a harmony with previous confused thoughts, in a kind of Drummond-tracery of conviction. Everything seemed clearer.

But I couldn't have got very fit; and as half-cured bodies were, very conscientiously, not just dumped back in England, I found myself once more on a train to a hill-station somewhere up in Bombay province.

I remember the heat of the railway carriage burning my thighs, and looking out of the window wondering whether I really wanted another session of hanging around the fringe of hill-station society. I was fed up with being a sort of pale shadow, living on other people's politeness, smiling when smiled at, accepting the unspoken comparison in passing Wrens' eyes.

But, deeper than that, I wanted to think - or rather to allow my thoughts to come. I wanted to explore this thing that was pervading my mind. I felt that here was the crunch; either Drummond or the facile values of the hill-station. In those passing fields, undistracted, I could perhaps work out its promise. Perhaps there were further absorbing discoveries - theme answering theme in areas of past gropings, where thoughts had criss-crossed unrelatedly and had always sunk without trace. I was tempted by a new feeling that got right where I needed it and stayed there, inviting peace, perhaps embracing new horizons.

The lure of the holiday hill-station - for it did have a lure - seemed to invite the muddle of restlessly harbouring secret thoughts at cross-purposes with the 'socializing' environment. That way lay headwinds. Thinking one way, perhaps when alone at night; then a helpless idler by day, not sure whether to drink too much, or a bit much ... I might never find the physics of the real Me, might never be able to love it over to others, like Drummond.

The little train had left the plains and began climbing slowly into the hills. It started to slow down, drawing into a wayside village station. Impulse teetered into action, and I grabbed my luggage and disembarked, not knowing where the sudden decision had landed me.

There was, I gathered, a small Parsee hotel a little way up the hillside. As I started up this path, a young, long-haired urchin begged to carry my case. As he walked alongside me, I noticed that he was a cripple. Was it three annas he wanted? A pitiful amount; the case was half as big as himself. Not quite starving I thought, though nearly.

Suddenly it hit me. Meekness and lowliness. Do it unto one of these my little ones and you do it unto me. What wretched, conceited, blind fools we Sahibs were. Poor little bastard!

I looked at him. His questing eyes were sad, imploring me. I suppose that only hunger would drive him to want to try and drag that awkward case up that hill. It was all too much, and all too simple. I asked him if he lived up there. "Yes, Sahib, just over the hill." A huge wave of conviction excited me, thrilling and warming me - the *me* that had just been sitting in half-rebellious dejection in the train. Incredulously he twigged that I would carry the case *and* give him a piggy-back up the hill at the same time. No great feat really, but I suppose that an encounter with kindness of any sort, even in his village, was none too common. And I had come to life again.

I put him down outside the little 'hotel'.

"Tumara nam kya hai?" ['What is your name?']

"Lakshiman".

I supposed he was about ten years old. He disappeared into the hotel, feeling I think that this Sahib was somehow in his charge. The hotel consisted of a group of outlying chalets around a central dining-room affair. Primitive enough, but an arrangement much to my taste. I was happy. Here I could secretly work out in practice the promise of those new ideas. I felt released from the old chrysalis - that questioning: "What do you think of me?" skinny as I was, which got into the front of my mind in every little relationship.

The Two Himalayas

The first meal was rather on the big side, certainly for my delicate tummy, and I think some was smuggled into young Lakshiman's! After a day or two the meals, mostly served in my chalet, grew bigger and bigger as more and more of them got smuggled up the hill to Lakshiman's village. There were lots of winks and grins, and perhaps, strictly speaking, it wasn't quite honest, but the simple fact is that the whole village might almost have lived on the nine rupees I paid daily for my meals.

The village belonged, I understood, to the 'backward classes', not long since referred to (officially, I believe) as the 'criminal tribes', members of the lowest Hindu caste, possibly below even that. Lakshiman was an orphan and he followed me, though not precociously, like a faithful dog. In return - for I enjoyed his company - I let him and his young friend sleep on the floor of my chalet, a relative luxury. The quite stunning thing was that nothing, just nothing, disappeared. I left the door unlocked day and night, and left my fountain-pen, and all sorts of relative valuables, lying around.

During the four or five weeks of my stay, one of the village boys was put into the local police cell for wearing a pair of shorts I had given him - the police assumption being that anything they saw them with had *ipso facto* been stolen. This happened again, and again I had to rescue one of them. No doubt they did, traditionally, live by their wits. As far as I could see they just had to.

Shortly after my arrival I gave the village boys a few bars of soap - partly because they seemed never to have seen it at work on anything. I still see their grins, covered in lather in the stream, all thoroughly enjoying it and, I shouldn't wonder, half expecting to be arrested for having it anyway.

I had managed to cut Lakshiman's hair, which had been hanging right down to his waist - in fact I hadn't been at all sure at first sight whether he was boy or girl. Apparently nobody bothered, as he was an orphan. He didn't fuss about my 'styling', and when I next turned up at the village nobody else did either. I went up there quite often, and in spite of their poverty and filth they managed to make me feel welcome and at ease.

I was particularly distressed by the misery of one tiny child who was covered - literally encrusted all over - with some horrid kind of eczema. The young mother, I suppose only about eighteen years old, spoke a fair bit of English, which mystified me. She was in fact quite attractive, and something about her suggested that she had come from a better, even a well-bred background; but she must have been cut off from that quite early on, because her ignorance was appalling, presumably just like everyone else in the village.

The little baby would crawl about on the mud floor of the hut, clambering over one or two other children. Infection? Never heard of anything like that, as far as I could make out. This was too much, and I offered to meet her with the baby outside the local doctor's surgery down in the village, saying that I would pay the bill. She said she would be there, but she didn't turn up. Oh well, I thought, women in this sort of village ... how can you tell what they are really thinking? Try again. Apologizing in a funny sort of way, half embarrassed, half evasive, she did the same again.

This was getting disappointing and a bit hurtful, and there was room for irritation. But it struck me that we were back somehow to the meekness issue. My being let down wasn't to matter, her embarrassment might. And thereby, and I'm sure only thereby, she suddenly blurted out something about the "green goddess" - their name for the

The Two Himalayas

disease - some kind of punishment from the vivid imaginings of their persecuted minds. It is hard to imagine the awful horrors of that fear-ridden mental world, but this experience shows how real they are and what a grip they have got.

I hope and suppose that some of the fear must have gone, because she did in fact come next day to the doctor with me. Not unnaturally, he attributed the eczema to a lack of protein in her diet, which was reflected in a deficiency in her milk. If I did a little about that, it couldn't have had much effect by then.

Lakshiman would accompany me to the local market on shopping expeditions. He had quite a little personality and saw to it that I only paid the local price, not some (very) inflated Sahib's price. Through one or other of these trips I met Tagore. I don't remember his other name, but he was, I believe, the self-disinherited heir of Rabindranath Tagore, India's greatest poet. He kept a little shop in the village, having chosen to opt out of his family wealth rather than conform to the egocentric customs of an autocratic life in such a poor country - ways he found particularly amoral, and unattractive.

He was a most impressive figure, with a flowing white beard carried by the dignity of his powerful shoulders and big - unusually big - frame. In his quiet, kindly manner he served rice and grain and little odds and ends at prices that annoyed the other shop-keepers, but seemingly with no intention of grabbing the market and getting on in a big way.

I accompanied him one day to a little railway station some distance away, where he wanted to submit a tender for the distribution of Government-supplied rice to the coolies working on a railway extension. The rival contractors all met in some kind of waiting-room. I waited outside but the meeting grew noisier and noisier and eventually broke up in disarray.

It seemed that the normal practice was first to meet and agree upon the proportion of rice that the contractors would filch for their own ends, and for them all to tender on that understanding. The noisy disarray was because Tagore had tried to break up the racket somehow or other, hoping to tender in such a way that the coolies did not lose this third of their ration.

Tagore longed to visit England, and I would dearly have loved to take him. In fact I tried, but the paperwork and regulations became so daunting that I gave up the idea. We had wanted to take Lakshiman. Sadly it all fell through.

I was very happy in that village and thought very seriously of staying on somewhere there instead of coming back to England. But that again was almost impracticable, if not completely so.

I left Lakshiman one day. Poor little devil, he was so down. Maybe I tried to give him some sort of start in life. I know that I worried about him. And it seemed all wrong to leave Tangore. I think they are both, indirectly, embedded in the reasons why I have written this book: because the time there with them crystallized something . . .
Perhaps I was lucky to have been able to find such a setting; to find, as it were, a sort of Galilean shore, calling me in its simplicity.

I have never, to the moment of writing this, felt so deeply amazed that Drummond's writing, or Drummond's approach, could be so luminous in a practical situation. He had got me spiritually agog, and then, it seemed, brought all my thoughts and feelings into a symphony as I worked it out. And it had all started from my reading the words, 'Take my yoke, learn of me'.

I have not mentioned a strange encounter, the significance of which escaped me at the time, but which really blew my mind sometime later in England.

I had met one of the other two or three hotel guests down by the small secluded swimming pool. An Englishman had suddenly turned up with a lady companion whose bathing-dress consisted, intriguingly, of about four or five handkerchiefs and a few (I suppose) safety-pins. I had had to kept submerged and rather away from her for the simple reason that I was wearing no handkerchiefs. He twigged and she didn't. Apart from the villagers, who never came near, I hadn't thought there was a female for miles. This resulted in the odd casual talk across the dining room.

After a few days he surprised me by asking if I was looking for the 'Good Life'. I hesitatingly replied that, in a way, I supposed I was. "So then", he said, "come with me tomorrow, and I shall take you to a strange place in the mountains where one Shri Meher Baba, a wise, wise old man presides over the lives of others, younger seekers. They come," he continued, "from great distances, and a taxi picks them up and drives them free for the seventy odd remote miles from the railway station. There they stay for a while, until one day Baba calls them and tells them finally how he feels their lives can be fulfilled."

Some, he explained, go to help the poor, some to teach, and nearly all entrust their lives to this man whose intriguing title means, I was told, 'Holy Father Mercy Baby'.

Later, I was to learn that this title means something not wrapped up in quite such Eastern-sounding mystery, something more like 'Holy Man of Learning' or 'Thrice Holy Man of Profoundest Learning' - the language stretching out the thought.

Be that as it may, I was fascinated enough to pack my case the next morning, and to appear at his chalet ready for the journey. But when I stepped inside I found the whole place full of weird gadgets, 'motifs' before their time, pieces of wire stuck into corks and moving in the draught, and so on. I thought he was not quite right in his head, so I excused myself and he went away alone.

Some time later, just before leaving for England I was, as usual, in some kind of transit hospital. By a strange coincidence it turned out that the young Indian in the bed opposite was the Shri Meher Baba's nephew. So he did exist, and according to the young man it was all genuine. He himself had been there, but had been advised by his uncle that he was not really cut out for such things. I felt not a little sorry that I had turned back.

Yet the Shri's message had been waiting here for me all the time, because very soon after I landed back in England, I innocently went to the cinema to see the film of *The Razor's Edge* - the Somerset Maugham story I referred to earlier - and found myself almost indentifying with young Larry, who went from wealth to attic, out to the Himalayas, and then dredged the waterfront world to find the Sophie of his earlier, casual days. It was just a film, but of something real. And it all fell, not quite sadly, into place in my life. Larry had gone as a pilgrim to the distant Shri Meher Baba, where I had nearly gone, and had taken the plunge - which I had turned back from at the most critical turn of my life.

The Two Himalayas

So, it was left to this film to tell me what the Shri Meher Baba would I know have told me - that seemed to have hounded me, which could get me bemused and lost - that a sweeping urge to give oneself for something is embedded in Man's Ultimate, and over the next hill, as it were, to the path to rest. My 'Razor's Edge' feeling.

And maybe I'd have mixed this anyway with a lovely female. I have said something about this tendency before, when reflecting upon my youth and the aching and intertwining needs of that time. It went very deep, dredging into where my Being came from; and my thoughts hesitated, uncertain of themselves, wondering, half-poised over Ireland. Perhaps, in my heart, Sophie and Molly were all mixed up?

I wonder why I never wrote. Too many tears, now all wiped away? Probably.

5

Whitsun 1946 was an experience; blue skies; lying out on the heath, the long years of war over. My body had decided to live. Nature and months of rest in England were building it up again. Everything seemed unreal. I ruminated often, mostly late at night, by our little local river.

I wandered nostalgically over to the stables where, it seemed ages ago, that wild girl of the forest rides had struck me down and bored right into my heart. I was still nursing the unrequited feeling, but Tom talked mostly of passing things. We had both turned up there out of the blue that day, and he remembered me as a chap who liked to sail without an engine. As he spoke, I looked out of the window and scarcely listened.

It was a whole war since we had met, Tom and I. We hadn't really been friends, but the deadly boredom of my pre-war years had once been broken by an invitation to an evening in his house in Kensington - Kensington, with its awesome, magical connections to the wider world. I can remember that evening to this very day, such was the desolate nothingness of those earlier years, and I remembered it as he talked of trivial things. I had been secretly astonished by the record-player and the subdued, indirect lighting of the new-fangled wall-lamps; but perhaps Tom hadn't realized how strange it all was to me.

His father, a director of some canal or other, was looking for skippers for their new barges. I had become a lot less intrigued when it turned out that they were to be diesels. We had chatted afterwards, about canals and bringing in diesels, and the old horse-drawn veterans; hence somehow Tom had remembered my thing against engines.

I snapped out of my nostalgia when he began to speak of my skippering his boat - or rather the boat *Foxglove* that he was itching to buy, and which I felt he was a bit nervous about.

He wanted to try his luck with only a tiny engine: that's how the boat was, and that's how he wanted it. I quickly discovered that that was about all we had in common. He was an infuriating perfectionist, full of mathematical theories about aerodynamics that were going to revolutionize sailing (always provided you didn't miss the tide over one or other of his fiddling perfections), and quite unable to see that a tall triangular sail could not be folded neatly into four equal parts on the boom till next weekend. And all this as it was getting darker and darker, colder and colder, and me hungrier and hungrier.

Yet he was kind-hearted, a complete atheist, and a nudist (the first I had ever met or maybe even heard of), and he had an equally perfectionist girl-friend. I recall how, years later, he came in from his rose garden looking cross and puzzled, moaning about women. Apparently the two of them had been sunbathing, and he had rested his hand on her left side, somewhat above the waist; whereupon she had complained that it wasn't half-past five yet. The significance of that escaped me, as it had escaped him also. It seemed, though, that you mustn't do even a nice little thing like that before half-past five because it would keep the sun off her, maybe unbalance her two sides by half a roentgen. He didn't know, and neither did I. Perhaps you had better make a note of it, too.

That garden, which I later grew to know so well - way out in the country - was beautiful beyond anything I have ever seen: a sloping hillside where three streams met; a riot of exotic colour that was, in the only words I could ever find, ridiculously beautiful - with bamboo thickets and other semi-tropical, luxuriant plants quite unknown to me.

But to get back to *Foxglove*. The previous owner had, for some reason, left her at Larne, in Northern Ireland. Tom, who had an eye for a bargain and a weakness for Ireland, plus lots of cash to overcome whatever difficulties might arise in getting her sailed back home. Thinking of Molly, I confessed to more than a weakness.

'Razor's Edge', that theme again: nostalgia and sex, a bad conscience - perhaps a feeling that my spiritual arrogance had rebuffed a natural, pure and virgin love - these all went over and over in my mind as we took train and ferry and put out to sea. I say, quite simply, we 'put out to sea', but rarely is it that easy. I rather think that a whacking great tip to the locals came into the proceedings somewhere, because it didn't take too long at all. I had six weeks free, anyway.

In the background, comforting me, there was by now the assurance of a little flesh on my bones, a little blood in my veins. Reassuring, not only for risking long hours out at sea, but because I had never really resolved the one about showing her my skinny frame - a 'Me that had to be at rest from my unfleshed bones', as it were.

Fresh air; a settled steady breeze. I stared out to sea and shook my head at that unwelcome, skinnier, ghost from the past. I suppose that somewhere I felt thankful.

Tom knew nothing of all this, of course. He knew very little, either, of the painful gaps in my navigational abilities. Aware of this, I wondered if he was trusting some half-pretender who was using him. I fell asleep in the sun on the deck, confused by all these irreconcilable thoughts. And when, finally, the wind blew us into the Foyle, the tide helping a little, and we dropped anchor, the flapping of the sails took over from my brain, and I felt better.

The river was of course very quiet, but as I lay in my bunk I almost expected to hear the thrashing of a destroyer, as I would have done in the old days. I think it could have run us down; as far as I was concerned it could almost do just that! In my book they had earned the freedom which they had saved for all of us. Not they alone, of course, but I remembered the threat of all those sinkings . . .

A wave of nostalgia left me with the feeling that I had never let them down - I don't think that I had. And that put all the sorry story of the rest of it into sudden perspective. It was all a punishment, which I had taken - fouling all the red tape - in order not to let *them* down. I fell off to sleep, happy - with that new happiness added to those other, quieter, happinesses that had I suppose come out of it all.

Even when I awoke there was a new confidence, a feeling that I could look back with a new smile at that day when I had found my way through the streets of Portsmouth; a smile that I had lost back in Bombay somewhere, in all the muddle.

The Two Himalayas

We stayed there a day or two, both pretty tired, and myself not averse to just being there, on that now friendly river. Then we left for Lough Swilly where we spent another day. Finally, we chanced our luck onwards to one of the smaller inlets shown on the chart (about which I, at least, had secret thoughts). I began to feel that I hadn't let Tom down either, that I wasn't too much of a fraud.

But by then we had begun to get the measure of the 'comfort' aspect of the boat. Tom, at least, was coming round to the idea that her cabin accommodation was all right in small doses. She was probably unsinkable - built from beautiful timbers of the highest quality - but nobody had given a thought to those human frailties that saw cricks and cramps as unnecessary to enjoyment of the passing hour. The cabin top was a bit low (to avoid windage, I suppose). The three-quarter ton keel was rack-upable, and rack fore and aftable. (The overall sailing qualities, predictability of, come into this feature somewhere). The keel stuck up through the middle of the cabin table - in fact a giant, elephant-sized 'centre-plate', as in small dinghies. *Foxglove* worked out at about ten tons and was thirty-four foot long; so you can imagine the keel!

You could look down through the middle of this 'table' and see the water, as the slot for the keel to slide in was of course longer than the keel. This seemed OK for watching fishes, dropping fag-ends, or gazing down into the clear bottom of an Irish creek. But, as we were to find on the way into the last creek, the sea squirted up and hit the cabin roof when it started to pound: a sort of hydrodynamics that I don't quite follow, but maybe a bit like the water-pistols of our youth. Getting very wet and mistrustful of what could happen with untried designs, we found various blanking-off pieces just beginning to float, under one of the bunks.

The winching arrangements were both lethal and dismantlable - steel wires running along the top of the cabin table are not notably pleasant. The dismantled winch was too big and clumsy to stow properly, except in front of the teapot. Cooking facilities were almost non-existent, but I suppose that this was tolerable because there was no room to stow any pots or pans anyway.

To reduce draught in the water you pulled up the centre-plate, having then more or less cut the cabin in two - a vertical plank with a bit of table left on either side of it. You couldn't then sail, nor could you see across the cabin; nor in fact could you close the sliding cabin roof. Some kind of tent contraption had to be lashed into place for the night. The only advantage I was ever to discover in this arrangement was that I heard less of Tom's snoring, which would drive me to distraction. I remember stuffing my ears one night with a bit of soft cheese (the nearest thing we had to cotton-wool) and climbing hopefully back into my sleeping-bag. It seemed unsociable to leave for the nice, quiet, wide bunk up forward.

We parted for some two weeks or more, Tom hoping to find a professional crew to sail *Foxglove* home, and I, of course, made off on an old bicycle towards the little village where Molly lived - or at least used to live.

Fortunately I had grown a beard, and I got - unrecognized, I believe - as far as the tiny village shop. I had been getting progressively more worried about possibly embarrassing poor Molly by any sudden reappearance, but this was where luck stepped in. There she was, serving in the shop. I wondered what *that* meant - working to blot out some misery perhaps? I stood and watched her serve a small child with a toffee-apple.

Suddenly our eyes met, hers dark and questioning. I may have bought a toffee-apple, for all I know ...

Our eyes met again that evening, some way out of the village. I might have noticed that her darkish-brown hair was a little longer. I did see, once again, more than a simple country face - those subtler, intriguing lines of hidden character. Once again it arrested something in me.

The sudden gulf that had opened so rapidly between us on the last day or two of our earlier time together seemed unreal, something of the past, a mistake rumbling round in my conscience, some hurt she'd had to bear. But I realized that the gulf I had opened had left her stranded, with a desperate and insoluble worry about her belief in confession conflicting with the fact that confession had torn our love apart - a worry that must have cut far deeper into her than any apprehension of mine about priestly intervention, or behaviour-patterns and things like that.

To be fair to myself, though, I had feared a spiritual apartheid. The idea that unnatural confessions could be twisted out of natural love dug right back into the quagmire of my early youth; where the rigid sexual judgements of the time had forced me to my knees night after night; and each night's private confession had cleared the way for Nature, reacting from the feeling of released tension, sighing to act again. I had been still very tender and violent-thinking about not getting into another mess of that sort.

Perhaps that old ingrained fear had been all the more powerful to divide us that earlier time, destroying an idealistic dream-world. Never related in our minds, I was sure, to anything like the bed-world, where it might have been forgivingly different.

I had a lot to be forgiven for. The mental position I had adopted had been just as uncompromising as that I had feared in the priest - and maybe unjustly.

Something of her lonely distress was reflected in the sadness that came and went that evening amongst the forgiveness in her eyes. It was nearing midsummer and I had hours in which to watch her face. Now and again she seemed to be asking me something - something very near the core of some struggle she had not quite resolved.

It wasn't until later that evening that I first caught a glimpse of that independent streak in her, thrilling to my youth: a brave little stirring from under the spiritual mould of village thinking.

She picked up a little stone and turned it aimlessly in her palm, and as if this helped her she threw it down and shyly, but purposefully, reached for my hand.

"Fred", she said, "Fred, but have you forgiven *me*?"

"Forgiven you?" But a glimmer of light and even a tiny hope was dawning as I said again - "Forgiven *you*? what for then?"

"I'm still in a struggle about it, and have been since the day I saw you go - or let you go, perhaps I should better say. I mean, wasn't it perhaps wrong of me to go to confession? I couldn't help it then, but when I found that it had hurt you - well, I'd begun to ... well to think maybe purity in God's eyes might be impure in man's - in some men's, and not in others. I mean, if nobody's hurt ... but I really don't know ... I have wondered whether even confession could make ... could sort of sully what is unhurtful and pure and natural - sort of God-made natural."

The confused thoughts came tumbling out, as no doubt they had tumbled in and out of her mind all that long time.

The Two Himalayas

"I think . . ." I started to say something, but checked my too-slick tongue. I knew that my hopes over our deepest problem lay in the survival of some at least of these thoughts.

I lay awake that night in restless, half-hopeful turmoil. Half-hoping that the old religious barrier was melting before this unsuspected independence. My sleepy mind galloped ahead. From somewhere or other the phrase came back to me, rather dreamily, melting more physical barriers, "She thought she would die of shame, but the shame died." Somehow the confessional faded and died, leaving her quite alone.

One or two verses from an old hymn - although generally I don't care too much for hymns - came from somewhere, enfolding her in my mind,

> O Sabbath rest by Galilee
> O Calm of hills above,
> Where Jesus knelt to share with Thee
> The silence of eternity,
> Interpreted by love . . .

God knows what the real connection was at first. The shame died . . . left alone . . .

Over and over again, all mixed up, very near to sleep. She thought she would die of shame, but the shame died . . . the silence of eternity, interpreted by love . . . interpreted by love. . . until the hymn interpreted how the shame might die, and I fell asleep.

There must have been staring at the ceiling, too, in her little cottage bedroom. What confused candle-lit background there may have been to her budding independence of spirit I do not know.

The next day was a Saturday, and my thoughts turned towards *Foxglove*, anchored in a lonely creek a few miles away. I wondered whether she would speak of it. I longed for her to mention it; perhaps it would mean . . . ? Although I was not on endless holiday, this was something which Molly would, I thought, have to find her own way to. Otherwise I could see the confessional feeding on some lurking guilt-reaction. No! The candle would have to flicker and be blown out calmly, without lurking corners of doubts. I was dying to drop a hint, but took refuge in somehow guiding the conversation to India, not realizing that a blank stare about these Indian things would have been, deep down, worse than any rejection of my hopes about the boat.

A blank stare, no sign of inward comprehension, would have been worse even than another confessional. Apart from one or two soundings, as it were, I had all along been funking the subject. So I changed the subject again, instinctively feeling an aversion to too much India bullshit - too much talk of far-away places. I was hungry for love, but not a hollow, evaporating romance.

About midday she suddenly said, "Fred, couldn't we go to your boat? Can't we get there somehow?" As she looked at me I lowered my eyes in case she read my thoughts. Something told me that she had blown the candle out that previous night. My heart raced, but I didn't want her to guess. Better she cuddle her own secret for the journey; then she could cuddle me more spontaneously on the boat - better not risk an accumulation of conversational doubt-reactions whilst waiting for the bus or something.

The Two Himalayas

But the bus or something turned out to be the one and only local taxi, and during the ride I was instinctively happy (perhaps also crafty enough?) to leave her cuddling her secret while I talked to the driver about dragging anchors and pubs, creeks and little villages. We two just held hands. I wondered if she sensed?

Untying the dinghy, rowing out, and such-like, were fun; light-hearted, reassuring things. We scrambled aboard. I noticed how lithe and agile she was; not at all at a loss as some girls tend to be. There seemed to be a natural intelligence behind the way she took it all in, a confidence behind the way she contrived a cup of tea in the strange domestic set-up.

There was some other confidence there too, behind all this. My uncertain nerves needed that; I was dreading having to be gallant and take her home. But, as the afternoon became evening, that area of uncertainty receded, mutually avoided. And so, at last, our eyes exchanged secrets. I had to be gallant in another way. Impelled no doubt by nature but also by some sense, I think, of spiritual affinity, she relaxed into my arms.

Yes, I had to be gallant in another way, to have such trust heaped upon me by a pure, untouched girl, who had only ever thought to trust her inhibitions, and now completely trusted me. I sensed that it was a complete trust and I struggled with tumbling thoughts. In a way our past seemed to have telescoped into this evening, tonight; but that shy ultimate, that seemed to be resting in her eyes, made me vow some secret vow that it should be a longer path to what we both knew was unspoken. I think we both knew and did not know doubt, but did not know how this could be.

It was as if our absence had been a courtship, had produced an instinct surer than any inbred instinct of slow betrothal. Any doubts I might have had about our not having had the usual days, weeks, months perhaps, of conventional dalliance were pushed aside by the whispered secret that she'd "nearly become a nun". I wondered whether or how much that could have related to my having gone away, and perhaps a shadow of conceit was there for a time. But the idea was pure and lovable, and I also loved her more because of that "nearly", so that perhaps the conceit dissolved away.

She was in my arms; and I saw a young novitiate kneeling with confused thoughts about confessions; of impurity where "there was none" - could any other nuns have thought like that? I was thinking of her 'God-made natural'. I didn't yet know that Drummond had melted my old idealistic "never-never" ideas away.

She looked up and said "I wanted to become a nun, because of you; but because I can't just brush aside your ideas, I couldn't. See?"

Poor girl. I thought back, too, for a minute about my idea of sending her to an abbess. I wondered what of yesterday's confused independence would have survived; and frightened, I shut that door of thought.

I confess, too, that her 'God-made natural', although sacrosanct I was sure in one man, seemed no thing for a timid girl. I held her tight, excited and protecting. Protecting her from being a troubled nun indeed, but a timid girl? I wasn't quite so sure. "Impure in some men's", but "not in others". I wondered where that could lead. Naughtily, I wanted to protect her all the way, I didn't know that sort of purity, where it led.

The Two Himalayas

As if to tear my thoughts apart, but really impaling me on the better of them, it suddenly dawned upon me that she had never seen a man before. God! I thought, that can only be done this trusting way - and God! her trust was very brave. I held her close, the nun, her last bedside candle, thinking.

What torrent of thoughts assailed her in the fading light I do not know. That last journey to the boat must in a way have seemed awful, final. But we both held hands as if the whole of life was being poured into this sudden day. Except that it was meant to be; all the first chords of some longer sacrament were playing on our sensitivities. But only the first chords, because of our sensitivities.

I had never really looked at that forward cabin. It had a bunk on the starboard side and a narrow table built against the opposite curving bow.

By hardly spoken mutual consent I lit the lamp there, and saw that two little bunches of wild agrimony and a nice bit of female tidiness had appeared. She sat down there on the edge of the bunk, just breathing the time away. Which seemed to mean - some telepathy - that I was right, she had never seen. I listened to the water lapping and wondered what it was trying to say.

Quite a long time passed. I had a feeling that she had passed through any valley of doubt, but that she wanted stillness to take her through a long tunnel of her own. So I sat there quietly listening to the water and, I suppose, to the happiness lapping at my soul.

I began to feel that she had mentally dressed herself in white symbolic robes; that fate wasn't very far away, and that I was to find some little ceremony for taking them, all of them, slowly away.

The lamp seemed wrong.

"Would you rather it was a candle?" I said; "Just a little one?"

She smiled shyly and, I thought, gratefully.

I took my nice clean shirt off as the little flame settled down, and I stood beside her stroking her neck. With a 'be kind to me' look, she closed her eyes and sat there waiting. As I stood there with fate, it seemed that the outline of her bosom was covering the waiting, each breath suspending the betrothal: and so I gently took her top off. She looked down at herself, so intensely deserted by her years of modesty; and then, halfway between pleasure and the verge of crying, she smiled bravely up at me.

"What about the poor candle, love?" I asked, feeling that we were sharing this gentleness. She knew what I meant. Her eyes already spoke the language of feeling very naked for the first time. But she reached along the little cabin table for the matches. Then across me for another candle. We watched her breasts moving for the first time.

The new light straightened out and settled, and she was still too. We were sitting there quiet. Then I heard her say, "I must be careful not to fluff it out, mustn't I?" And with a brave, almost saucy, smile she stood up, very close in the little cabin, holding on to my shoulder - her clothes all gone.

For some unaccountable reason her shyness seemed to lift for a minute. Then she sat down beside me again, waiting - now timid, and yet bravely not clinging - she naked and me not. Something about the way she held her head told me that this candle would complete our journey into nakedness, into all the thoughts and initial feelings of being virginally bare together.

The Two Himalayas

The same candle would light up her eyes; perhaps see her shrink into herself; or perhaps see her breathing only her want for me. I said to her, all my instincts wanting to help her, "You're lovely." Which she was.

"Am I?" she smiled, and she lowered her eyes. She glanced at me once or twice, then suddenly knelt down in front of me, her hair falling over her face, her hands crossed just below her neck betraying a surge of modesty. I had the feeling that she was dedicating herself in some way, and we were both still and quiet. She was there, kneeling, for a long time, her hands now gently on my knees.

Then to my surprise her hand reached out behind her and moved the candle so that it shone more upon me. She raised her head, shook her hair back, and as it were fully and frontally looked me in the face. Maybe she knew that her random little curls didn't really obscure the little valley, but her eyes stayed still as mine very slowly carressed her body. Her faint smile told me she didn't mind; then she glanced over her shoulder at the candle and back to my face, a little wide-eyed now. She took her hands from my knees, even involuntarily leaving them slightly more apart; from which I knew that my clothes were to be gone while that candle still burnt. I knew, too, that it was to burn on a little while before she would look down and learn, from me, about men.

As unobtrusively as I could, I did as I think I was bid, awkwardly; somehow, still sitting. But she stayed spiritually concerned, not looking. It was as if her thoughts were sublimating an intense feminine wondering curiosity. She couldn't have known how very much, paradoxically, that drew me to her.

Suddenly I saw from the little changes playing in her face that there was a different kind of thought, and as if excusing herself she said, "I'm trying not to be silly, Fred", still not lowering her eyes. "I've thought about your sort of freedom; I want to be the same. To me, it means that I can kiss you there. It's a dream I've had, often, thinking that I must be innocent for you, yet grown up for you - you've respected my childish ways long enough." She paused.

I was shattered, knowing that she had never seen a man. Having never even thought of such a thing, I found it almost intolerably exciting. Those vague dreams. Floating as it were in half-shy ignorance; her trust and femininity homing just on to "me", just "there".

"No, Fred, I don't really know, but that's the way I want it".

And still not knowing, she slowly bowed her head; I thought she hesitated, her fingers tightening a little on my thigh; my brain tightening between excitement and care. A distressed little moment started an unfinished whisper but which somehow changed into a half-swallowed "It's alright". Then I felt her relax; her fingers loosened, a few heart-beats, and I felt her kiss me. Somehow I knew that she *was* all right and that my delayed, stifled apology already belonged to the past.

I remember an almost childish look in the face that looked up at me; serenity came and went; and came back again, I think because maturity was there as well. But I knew she was wondering at me, and was also wondering at herself.

But I had been far from serene, caught up in a struggling balance. Half deliriously I realized that she needed to go on - to verify something, as if some sacrament had yet to be completed, because with a smile as of far-off things, she bent again. It seemed to me that time was suspended by those gentle lips, exploring gentleness. Somewhere in that time a wisp of hair fell, misbehaving, and wisped slowly, differently. And the struggle not to inflict something on her innocence increased in intensity so that, fearing that runaway kind of more than pleasure, I sought, distraction.

There is a limbo in the brain, it seems, for when we require anaesthetizing some little doggerel, or some words - quite irrelevant - stored there by past repetition will quite unaccountably come to our aid.

From such corner of the brain there came a silly little rhyme. Hopefully I could play with fire, keeping the excitement and recite this to myself to stem the flood of unmanageable, hazardous, triggering thoughts. It was a stupid mnemonic, an aid for remembering the bridges over the Thames. No! that was wrong. My brain groped elsewhere, tearing itself painfully away from the young body, but tearing itself away for her sake. I found myself mechanically reciting those old familiar words 'Do it unto one of these my little ones, and you do it unto Me'. That was, more appropriately, from somewhere deep inside me. But it seemed a little blasphemous in my rather fevered state. No, not that . . . Then at last came my recent discovery - the spiritual chemical that had been promising such a magic touch to my quieter moments, and had been in and out of my mind so often: 'If you are willing to do the will of God, then . . . you will know of the doctrine, that it is of God.'

I didn't realize it, but it was that faded little book again. That vague saying which had been tantalizing me with obscure meaning. It retraced itself through my mind, lending something to me. It looked down from the unknown to this moment I had never known, and gave me what I knew was of God. Timelessly it gave a serenity to the loveliness of my need; a serenity that transposed itself to mean that I must be willing, for her sake, not to tinker with those tempting thoughts. And the wonderful but racking temptation sublimated into a strange hyper-erotic peace. The desperate thinking stilled, suspended also in timeless willingness.

The little cabin scene, the candle shining and flickering on the curving young back. The charisma of those last words, 'You will know of the doctrine', illuminating the way to keep her safe and me intensed - the way hidden in willingness for her sake, and not in just plugging sex as it leaked from the brain.

She looked up at me once or twice, and I sensed that whatever thoughts she had none were about time or hurrying. So that my little bits of waiting were not spoiled and shadowed by the imminence of an end, rather did each minute go only to intensify the next . . .

But at last, increasingly wondering, I saw her face rise, her lips pursed in that endearing saucy little way that matches a tiny wink. What better than that I should wink back, slowly shaking my head, released but captivated. That mixture of childishness, maturity and sauciness had completely got me.

I groaned and muttered my way back to normality, but that could never be the same again. That night we slept together, but she knew that I would ask no more. I would rather have stayed awake all night, which I nearly did, because the gentle rocking of the boat was too wonderful to stand; but exhausted, partly with happiness, I fell at last asleep.

I was delighted and grateful that there was no reaction the next day. For some reason we were both a little shy about the closeness of the boat. Now and again I caught her wrapped in thought - contented thought it seemed. Now and again she had a proud look

The Two Himalayas

as if she had outdared and outmanoeuvred something that everyone else had feared, as if some unique act of faith had been justified - as indeed it had.

We were lucky too. Her holiday from the village shop - wangled I expect - the first holiday of her life, meant that we would be free to walk, and row, and idle our time in the little deserted inlet where *Foxglove* was anchored. The Sunday passed quietly by, ending shyly too. Monday was hot, and we were looking down into the water, cool and clear and inviting on that hot afternoon. Suddenly the mental consummation of that first night reflected itself with delicious daylight daring. "I can't swim," she said, "row me over there, it looks shallow. Let's try."

Again with that saucy look, she made it clear what she dared, as in a daze I watched her throw her clothes into the cabin and start untying the dinghy. Almost meekly, quite stunned, I followed suit. I was glad, for her sake, that the surprise stifled any other possible reaction - an indefinable, diffusing help. In any case, I was preoccupied with gazing out to sea, secretly trying to stop shivering with the supressed excitement of this dream-like idyll. And it was exciting, rivetted down by convention as we were - even 'topless' an unknown word - so unreal was this beach adventure. I think I remembered instinctively the remorseless mockery of that not-quite swim years before with that not-so nervous 'Anorexia Nervosa', the saddening and galling aftereffects: the frustration somehow of the Me, that special desert-island Me, left sitting - both ways - symbolically on the cliffs.

I may have tried to say some word or other, and she probably replied. Then turning round to me, blinking, I thought, she stopped half-way through a sentence; looked up to see if I had noticed the passage of her eyes, the passing flicker of her brow, then involuntarily scratched her head, just quickly, in a childish sort of way. I caught the profile of an involuntary, elfish smile that was not really for me - but perhaps about me, about Nature's funny wont. But it told me of the indefinable, relaxing friendliness of this nice unurgent picture of man - nearer those cottage dreams in which "she was growing up for me" I supposed; and I was happy. What a wonderful girl!

There's nothing like a swim together to bring a sense of proportion and I was grateful for the quiet undertow, as it were, in our relations, balancing the surging tides of the night. I could hardly believe that such a spontaneously happy dream-world had come to me; the sandy inlet, the grace of a young body suddenly freed to splash about, abandoned to my watching as if my love were her clothes. I felt that the secrets of each tiny movement, each runlet of water off her surfacing body, were tiny presents for me.

We lay in the sun on the sand, and I drifted off to sleep. I woke, thinking of a cup of tea, and so we scrambled into the dinghy and I rowed her back - almost unaware that we were running a one-towel gauntlet to the boat. I would have given something to have been stronger, but that blemish on my manliness didn't seem to spoil her happiness. I just couldn't get over my delight at this little adventure. Release to happiness, I thought; together with compliments, I felt, of the gentlest tide - as if that came with the compliments of God.

I felt sure that in some way she had caught a spiritual link between us that had given her this rare kind of unsofistication, beside which *sophistication* in the accepted sense would have been a thin mockery. And the cosiest of thrills, and intoxication of life at its best, enthralled me completely - quietly, but completely.

The Two Himalayas

As I pulled alongside *Foxglove* the idyll clouded. I gazed at how high she lay above us. I'd had trouble getting myself aboard before. I suddenly realized that I didn't know what to do. I had forgotten in my excitement to hang a rope loop over the side to use as a stirrup. I tried to climb on board and failed, and unused to the skittish ways of dinghies, Molly nearly fell in. I just wasn't strong enough. The deck was too high above us. She tried herself, and failed, ending up back in the dinghy. We looked at each other. Something had to be done. I hadn't tried to help her - a silent struggle about the unspoken.

"I nearly fell in, trying to help *you*," I heard her say; and "the days of chivalry are over then?" And her eyes registered a half-joke, a half-truth - perhaps the whole truth - I just didn't know.

"You mean I'd better help *you* get up then?" I said, half to myself, my voice tangling with my thoughts. I offered her the towel, all we had, but she gave me a little kiss, "I wasn't thinking of being only half yours," she said, looking up at *Foxglove*. The towel dropped in the water, sinking symbolically away.

She put her foot where my free hand could hold it, my heart pounding while she twice tried to struggle up. We were both very much aware of one another as the first effort failed. But, eventually helped up, she rather gracelessly clambered on board. And you could almost hear the silence of those thoughts that went unspoken. Not least - I confess - I wanted to stop the world as if to listen to her mind.

But with a deep breath and an oblique preoccupied look, she recovered her dignity, and came to the rail to help me up. I got up as best I could, but Nature had taken over and Molly, smiling the while, had to change her way of helping me.

"Is *that* only because of what I said?" she quizzed, puckering up her face but not, I think, her mind, as at last I made the deck.

"I don't really know," I said lamely, and sat down awkwardly.

"You'll have to dry my hair before you get any tea," she said impishly, and sat down beside me. Her eyes were soft and gentle, bewitchingly untroubled. "Especially if it wasn't . . ." she said. I suppose those sort of tides are funny things, and she would not have been too aware of the unpredictable nature of their effect.

"For God's sake get me that tea . . ." Cheeky thing! I wonder why I remember that native wit? In that strange land of my 'funny tides'. Her eyes too were all mine, not half mine; and yet they teased, unafraid, over what she'd said.

The next day we decided to try and get over to her village. Transport could have been simpler, but eventually some local worthy lent us his trap and pony - half horse, half pony; eager, but trotting when it felt like it. I thought that probably the elderly stepmother who welcomed us, and with whom I had stayed years before, was a bit too old to register very much; but I gathered that her sense of romance was still alive and well. We had a griddle over the fire and it was all as it once had been.

The day after, we set off again with a sigh of relief back towards the boat - a thankfulness of life, albeit tempered by an awareness of problems, of impending decisions . . .

As we rounded a bend in the quiet little road, we saw the unmistakable figure of a priest. Obviously he'd had a puncture. With a mixture of pleasure and misgivings, I helped him lift his bicycle up on the cart. Apparently he was near enough to his outlying parishioners to finish the journey on foot, but too far from his home to push the bicycle all the way back.

We took him where he wanted to go, to find that local improvidence made mending a puncture an impossible task, so would we not care to take him back home and stay the night? I did not exactly relish the idea, but common courtesy, and the prospect of a dark ride back to return the horse and then find the boat, and all the rest, meant that we could hardly refuse. Poor Molly's thoughts must have been in a turmoil, although I soon gathered that he was only there as a locum on some kind of holiday - not the same man to whom Molly had confessed all that time ago. The old stepmother would have been romantically glad about locums, as indeed I was. A nice enough fellow, whom I could take to, but secretly wielding the power of the confessional: comfort to those who came, deeply-rooted unease to any who held back. I began to have qualms, and visions of that miserable old split right down the middle of our happiness.

He chatted pleasantly enough, but one had hardly to be very clever to know that underneath were more serious thoughts. He was trying to be kind, but that had subtle dangers for a sensitive girl. I could see Molly's beautiful innocence breaking up into self-doubts. I remembered that night long ago when she had run away from me, and had vanished frighteningly into the mist and fog of what looked to me like some kind of horrid peat-bog. I cursed my luck more and more as an uneasy supper played out its little game on each of us. I say we chatted, but I felt that all the problems of some far away future were inevitably focused on both of us as we played with evasion and truth and guessed how it all sounded. I longed to get back to our little trap, with its squeaking wheel, and perhaps even a mare in foal to kick the situation to pieces. And I was desperately sorry for Molly, with her uncertain table manners and her attempts to brave out the pressures that she was so unused to - the silent yet overpowering Mother Church watching and disconcerting her every groping thought.

The crafty idea entered my head that the rather tubby-looking cleric might not take unkindly to a drop of wine, which I very willingly fetched from the cart. Perhaps it gave me a little more confidence too, for without the familiarity created by a little alcohol, I don't think any of us could have braved the plunge into the religious no man's land that lay in our path. Molly and I both knew that it would crop up eventually, but during all those days I at least had been sweeping it under the carpet.

Pouring him out a large glass of wine I tried, as inoffensively as possible, to reminisce about the effect of India, without bullshitting about India: to draw his sympathies towards the story I have attempted to describe, and to touch upon one or two themes that I felt might be of common interest. I hoped, perhaps, that there was just the chance of reconciling him to the loss - maybe in one sense only - of one of his flock. But what about the boat? Well, heaven only knew about that; except that he'd had the decency not to ask where on earth we could have been going. I admired him for that.

I had the feeling that he would like to have travelled and been a missionary; that his instincts and sympathies were very much where mine lay. I felt that rather than be a village outpost of doctrine, he wanted to love people and help them. He was more and more kindly to Molly, and I began to sense that he just wanted to give her his blessing. I am sure it was by no means all the wine. I poured my drink on the aspidistra when he went to make the coffee, knowing that an alcoholic haze could let the promise of this moment slip through my fingers - through all our fingers - save perhaps those of Molly, who seemed more than a little overcome by the spell of the tall, imposing dining-room, and the aura of centuries of religion: the life-long reverence for the priesthood. She was after all a village girl, born and bred to the Mass and all that goes with it. Her conscience was at work in her restless eyes, and she must have been wondering how much the priest knew or guessed about the boat, otherwise 'where on earth had we been making for' would have been hinted at somehow or other. Perhaps some suggestion of promiscuity was being scarcely held at bay. Whether she consciously revisualized and reassessed her part in that evening, or her utter daring in the sunshine, I did not know. I said a little prayer for her, that she would not - but I saw her eyes close once or twice, and feared for her. And I feared for myself.

The moment came at last, which she was half wanting and half dreading, when we rose from the table. I too dreaded the effect upon her of a lonely night in this place. God alone could know what unhappy creature would see me in the morning, after a night of tearing herself apart in religious doubts and harrowing loyalties.

True enough, I hadn't long to wait. The priest had shown us to our rooms: two tiny little bedrooms. We all went downstairs again, Molly and I both instinctively putting off the final parting for the night. The priest had disappeared somewhere outside. Suddenly Molly made some little excuse and disappeared also, into the kitchen. Wondering, and fearing the worst, I waited. Our whole future seemed to be clouding over with trouble, the sort of trouble that nobody could sort out, and which I knew had broken over her in the kitchen.

I waited until I could wait no longer, and there I found that lovely, happy creature of the last few days sobbing her heart out. Great uncontrollable waves were wringing her and shaking her whole frame as she bent over the mantlepiece. She looked up at me in despair. I don't know what I said, or what she said. I wanted to take her away into the night, but there would be no future in that. Instinct told me that this scene would haunt us on and on and on; whatever we did, wherever we went, it would go with us. This religious thing had to be faced - yet facing it, there was no answer.

I suppose that this melancholy impasse must have drained both the hope and the ability to feel other than numb. She was drying her eyes when the priest appeared from outside, blinking in the light, carrying a lantern. He took in the scene in a few seconds and, I suspect, guessed some at least of its significance. I took to him more than ever, as he came quietly across the room looking kindly from one to the other of us.

"Father," I found myself saying; "Father, I think I had better do the confessing. I'm not a Catholic, and I've not been brought up to it like Molly has; but if there's anything to confess it's *me* that had better do it, because it's all my fault - if fault there is," I added. "Look, Father, I'd better - I *want* to - tell you the truth; I'm sure you must have been wondering. I broke away - left Molly - three or more years ago. Left her in the lurch because, if you want to know, I resented her going to confession about some little thing. I've got a feeling that it wasn't you she saw. Or maybe you don't remember, maybe you're not supposed to remember, I don't know about these things. But I tell you now that the fault was pride and bloody spiritual superiority - forgive my language, Father! - on my part. I'll confess to you about *that*. But I'll tell you . . ."

I saw Molly glance apprehensively at me.

"Yes, dear," turning to her, "I think it's best."

And then to the priest again: "I'll tell you that Molly and I love one another. That there's a boat in the creek and that that's where I've taken her and where she's stayed. That there's an awful lot of trust, a fantastic amount of trust, and there've been two naked bodies . . . Sorry dear," as I saw her colour violently, where the tears had run down, "that Adam and Eve may have swum, yes swum, and climbed about together, and that Molly here has knelt down by me by candlelight and kissed me; and that the serpent went away!"

Fortunately I had dreamed up that little metaphor a day or so back, and it returned to rescue me. Molly and I joined hands, protectively, although a desire for protection from this man was nowhere in our instincts.

"Molly", I said, "suppose in that old story that the serpent is all about shame; would you have said that he's been a bother?"

"No", she replied, addressing the priest. "No, Father, I'm sorry, but he hasn't. Is that wrong, Father?"

There was a pause.

"Some people", he said, "most everybody, can't seem to be happy without first thinking of the body as evil, and then getting all ashamed about it, then doing wrong - what they think's wrong - and then confessing it, and then feeling alright once again. Mostly over and over, all their lives. I'm taking confession first thing tomorrow. I somehow deduce that you two are living amongst the rarer things of this world. I deduce - I don't quite know how - that it is not yet as man and wife . . ."

Once again the poor girl blushed.

". . . and if that is so, then I shall pray for you indeed. But not . . ." and he put a fatherly hand on her shoulder, "not at confession, my dear, not at confession . . ."

Poor little Molly turned towards him with a flood of tears, different tears, melted into a smiling, shy gratitude.

"Even if there were nothing left but sanctification," he added, as if to reassure her.

What a man, I thought. Indeed, he seemed to be a very rare type of person, and at that moment a very happy one.

I still remember the old clock ticking quietly on the wall, as if to indicate that there was a new kind of time in which nobody could be the same again, in which it felt wrong to disturb the friendly quiet of this old kitchen, now somehow a new kitchen. It ticked away, undisturbed, until the man of God picked up his lantern and beckoned us to follow him out into the warm night. "Don't ever let me down," he said mysteriously; "God bless you, both of you." And he turned to go, leading the way.

It was getting late, and I didn't understand why we were leaving the house. He could perhaps have found trouble with the pony - an excuse to talk of priestly things. But we passed that in silence. Or was it the church? Had he intended all along to bring us there, sensing our problems? Lit it specially with candles - to give us our freedom? But the church was quite a step away, beyond all these outbuildings.

A few more puzzled steps and he suddenly turned, opening an old wooden door. I saw his face in the light of the lantern: quiet, but continuing his blessing; serene, reminding me with a sudden strange conviction, of that special serenity of Molly kneeling on that first evening. He shone his lantern through the doorway, politely letting Molly step inside. Without any hint of connivance, with no more smirkiness

The Two Himalayas

than the stars above us, he let us see the hay - pausing while we took in the old beams, and the wooden ladder leading up into the shadowy loft. He passed me the lantern, while we both tried to take in something of the meaning of this, to us, so profoundly unconventional, so profoundly moving, blessing.

I had to say something, and stammered "You're sure it'll be alright?'

"I'm sure," he said, "God bless, and if you don't mind, could you just disturb the two beds I showed you upstairs." A little connivance showed in his eyes, but their light was tender, a touch of humour, nothing more. This hit some chord in Molly's nature, perhaps several chords, and she gave him a little kiss. And another little kiss.

"I've a housekeeper," he explained, "a sweet old girl; but I musn't be tempting providence. So maybe you'd better shake any hay off the blankets if you want to bring them down here." He turned to go. Impulsively I shook his hand, and said he was a brick. He smiled, and said, "She'll be here about ten o'clock, not before." and, as if half to himself, perhaps to reassure his conscience, he muttered once more, "Don't ever let me down - and I don't mean about the village. I mean before God."

Both our hearts were full to overflowing; until I suppose sleep came to us, in the hay that seemed to know us better than any pillow.

In the morning Molly cooked him some breakfast, her eyes alive with affection for this man of God, this man who had so boldly hacked down her fears and worries, and lit up for her that same, adventurous little path that her own soul had chosen - whose lantern stopped the candle going out.

We rode unevenfully back to the boat, stopping once or twice to lead the horse or to gather sprays of those multi-coloured mosses that enrich this part of Ireland. I remember too the honeysuckle and the fuschia growing half-wild along the lanes nearer the tiny villages. No danger now, I thought to myself, of Molly disappearing into a swamp, as indeed another priest might have . . . and the ghastly thought of how it all might have ended sent a chill through my soul. I felt safer, far more secure than I had done previously. There was no longer that nagging worry, which had been rare enough; but now and then her simple femininity (if such there be) had been quietened by a sort of brown study that she would suddenly snap out of, but which it wasn't difficult to account for - and in a disturbing way.

There was one of those long-ago evenings by the peat fire in her cottage when, maybe for the first time, I had felt her cringe from some spectre that she'd suddenly disturbed lurking in romantic secret idling thoughts. She had been in a far-away mood, telling me how she'd heard of someone "in the family" who'd been named Tristan because he'd been born on the island of Tristan da Cunha, and that she'd like a boy to be named that, because it would make him want to travel and go and see it one day. She'd suddenly grown silent, as if the fire - and something else - had died. She leaned toward me, her head as near my shoulder as it had ever been those nights - then shrugged it somehow away - until it sought me properly, abandoning pretence. I had felt rather than heard something about "that can't never be", and had sensed that same religious spectre clawing at that same unallowable dream.

We had never mentioned it again - we never did - but now the trap and its eager pony seemed to sing a song of an allowable unallowable dream.

We rowed - or rather she tried to row - back to the boat, 'catching crabs' as they say, swamping me. Underneath it all more sure of herself - her shows of ill-humour were if anything pretend tantrums, more extemporized nagging, than the real thing. She was no longer so hung up about the passage of time.

Rowing the dinghy started it, some quiet remark about rowing. It made me idolize her. Maybe it was that same night, perhaps a little bit snugger than usual, feeling cosy and close because of what had happened.

She was arranging a few wild flowers from the journey.

"Dinghies are nasty things," she said.

"I rather like them," something made me say.

"I thought you might," she answered, saucily referring for the first time to our little contretemps of that sunny afternoon. The expression on her face said more than words.

"Like candles?" I encouraged her.

"You liked it, didn't you? I had to make you dry my hair, didn't I?" she continued, gaining confidence. The saucy look faded, until her eyes looked straight into mine, close and soft, nearing some inner goal. "I haven't forgotten it. I don't think I want to." She paused. "Do you know, Fred, it was like a precious little drop of time, there with you, down in the dinghy. Because I think it was Me, the real me, the real me over a longer time."

It was as if the significance to herself of that afternoon's spontaneous, unambiguous reaction, her fleeting immodesty, had to be recaptured, and reverified. Referred back to that candlelight.

"Or I could have been frozen there *against* my will; sort of paralysed, mixed up, even sort of sporting."

She said no more.

Time stood still again - intensely aware time - suspended somehow in this question.

Then, at last, she said quietly to herself, "Not really. That *was* me; the doubts came afterwards . . . they don't really belong. The candle knows that."

I thrilled to a growing sense of this chastity, even in immodesty, although whether there could ever be words for it I did not know.

A passing fishing-boat faded into the distance. And as if it had come to take the doubts away, she freed her clothing and it fell away - leaving her finding words that I didn't think could be found. Words that I shall always remember as part of her genius, of her shy genius for this thrilling chastity.

"Don't touch me, Fred; but I want my body to talk to you again, to tell you quietly that it doesn't mind, to talk to you slowly . . . not frozen against my will, but to be stilled in willingness. A nun, and yet not a nun; sometimes just still, if you like. I don't mind."

She left the lamp on, but lit a candle.

"In willingness," I thought; I must have used that word "willing".

And she lay there quietly for my looking - stretching herself for the first time; not quite still, because of shyness, but stretching the time, as she knew that my eyes were young and would be happy.

At last. "I love you," I said, kissing her toe, and teasing her with a smile. She laughed gently, and moved slightly for me - than which there could be nothing more personal, and to me nothing more wonderful, because her transcendental sacredness won, and a new calm took over. A hypererotic sublimation from willingness for each other's sake that I think was of God; a translucent magic for the whole of my life.

A shower fell upon the cabin roof; little drops, patterings of the outside world: and my mind, and that helpless child of the mind, were into a covenant, some kind of sacrament.

> Rove over me, sweet eye
> For I am captured.
> Nor ebb nor flow of shame;
> It is all sanctified.

Some kind of sanctification was accomplished; but consummation was far from my mind.

Both her village and the world seemed far away. She slept peacefully on those nights, and I was strangely content to leave it at that. But more than once I laid awake trying to think where this was all leading: trying to figure out the future. Where were we to go? England? Ireland? Now? Later? A thousand questions each begged an answer. There was that old idea of asking an abbess to somehow employ her for a time, because the question of 'table manners', however distasteful or patronizing, might become one of those very real sources of hidden unhappiness. For the time being, though, I didn't reckon we were doing too badly with the help of stupid remarks, exaggerated pantomimes of posh society, that sort of thing.

Only once did we actually get to talking about the future. It was after supper, towards the end of the week. She had gone quiet.

"What are you thinking about?", I ventured, hoping that I hadn't started us off on problems.

"Can't you guess?" She looked down, away from me.

"I reckon you're wondering whether I'm going to brew up another shit theory that'll . . ."

"Like you did those years ago?" she interrupted, sadly, but not I suppose too bitterly.

Was this the ghost of that confession business, or was it another worry?

"Not quite", she went on, in response to my 'guess', "at least not a religious theory. Our friend, Father what's-his-name, didn't seem to want to stir that all up, so why should we? At least, I hope we don't. No, Fred, it's more practical things. I'm miles behind you in so many of my villagey ways. I'd be a drag. You're burying all that, aren't you? I'd become a bore, a nice, kind, thoughtful *lump*, but a lump all the same. The girls down the road, and in the office, with all their fancy ways - I bet you'd soon get all restive, wondering what you've missed."

She paused sadly, and looked down, slightly to one side. I thought her eyes were filling, and her lips were trembling slightly.

"Yes, and after you'd taken me a few times," - and a large blush complicated the breaking sadness of her face - ". . . you aren't thinking about *after* that, are you? You aren't thinking about my small world getting smaller and smaller - maybe us not even having anything left to differ about? Because you would have got to the bottom of me, and you'd know it, and you'd be stuck. And then you'd be stuck. And I'd sense it, and get worse. And then you'd start fidgeting your feet, like people do when they're waiting for an evening to end, and you'd be all polite... Oh, Fred! And me not knowing even enough to know what I could do about it; and everyone thinking of me as the *bumpkin* - you know what I mean? Oh God! it worries me."

I had put my arms round her, and tried to comfort it all away; an optimist in love.

But that cloud drifted by; lighter for having been spoken about.

The days on *Foxglove* were drawing to their end: swimming, rowing about, idling, and so on. Fortunately, Tom had not reappeared. About three or four more nights, and it would be our last on this boat in this lonely creek.

The moon seemed to know this, watching us - riding almost full. Molly seemed a bit quiet - but unquiet; a little subdued, but restless. Since the evening I have described, things had by mutual consent been very low key; just gentle, thoughtful good-nights, demanding nothing, resting in love; caring.

But tonight she lay herself down, the nightie far away under the pillow. I knelt beside her, yielding to her strange mood... and found myself yielding to the latent thing of that passive week. My feelings broke over me.

Her breasts became my inner longing, curving gently for past sadness - the age-old longing of man for woman. I had thirsted too long, and it now welled up in me. Her hips were my desire, but somehow pure. She knew all this, and accepting it, sunk back on the bed. After a while gentle fingers pre-empted mine, her innocence gave her a freedom, a delightful caressing freedom - whereas my sort of innocence was more of a constraint on freedom. Then my hand, feeling strange, surprised itself. My eyes followed. Imperceptibly she arched her back; and now my lips drawn into one unimagined, answering bowing down; the next step in the path towards melting together.

Conventions had deserted her, what little she knew or didn't know, melting as dishonesty melts before honesty, leaving her with me, yet in a way alone - floating in a strange half-lit cave, afraid that her feelings might sweep her over some waterfall; yet with a hunger crying for an end to stillness.

Her legs moved, and I sensed that only shared memories of those two earlier, trust-filled nights were left, because she whispered something from some chord of memory, smiling it down from a tousled, candlelit head, tantalising yet reassuring - keeping us back, but drawing us on. Till at last she spoke to me with another tiny movement, and her head went on one side, and her hand forgot.

Somewhere in the background, I knew that India and now this Ireland were disentangling that earlier hopeless Me.

I think I may have prayed for her, wondering if any man could bear such trust, and possibly feeling grateful somewhere for the understanding shadows of that old hopeless Me.

Maybe I shouldn't say so, but I believe we were both endlessly grateful for this rather shy, rather 'spiritual' start.

The next day passed.

In truth I cannot remember much. I suspect we were both miles away and yet very close. We were both so lost in random thoughts of past and future, that we lost the dinghy; but I didn't really wonder *why* anybody should tie a silly knot like that! And

The Two Himalayas

the following day she lost her right shoe and, rather illogically, her left sock. I teased her that I didn't quite trust the mushrooms she picked - and am not sure that I did!

The morning had turned drizzly, and while I motored round to try to find the dinghy, she got herself into turning out the 'kitchen', where a disdain for keeping your hair dry, and all that usual sort of female thing, had got things pretty wet and nasty. Domesticalities were really sometimes beyond a joke in that oddly arranged cabin. She had managed to clean a fish, and the scales had stuck themselves everywhere with their primitive type of glue. Eventually having had enough, she had come out with a bucket and stood in the now pouring rain with a bit of soap; one of those patent 'baths' that, with her trusting nature, she had thought I "wouldn't mind" since several days of intermittent showers.

I noticed how this time, though, she stood as far away from me as possible, and when I turned my head in her direction she stuck her tongue out. I pondered quietly the strange ways of females.

Oddly enough though, the next and last day, when I stood in a shower she menacingly held a little feather. "Pity you haven't tidied up these feathers," she said, "I may have to tickle you". But she hit the cabin roof, so I never found out how innocent or how daring it was supposed to have been. I suspect she was going to tickle my balls, but banged her head instead.

We turned in early, cuddling quietly and thoughtfully, with sleep not far away. For some reason I was back to thinking of what she once had said: "I must be innocent for you, yet grown up for you". I wished she hadn't hit her head, she'd probably have been both. That feather could have been an erotic experience tantalizing beyond any dream of mine. For some sleepier reason I wondered how anybody standing there with a half-challenging feather could even think of themselves as likely to become a 'lump'. That's what she'd said the other day, wasn't it? "A bore, a nice, kind, thoughtful lump." I smiled. I wasn't very worried, and sleep came nearer. Perhaps there was a bit of a lump on her head? I cuddled her closer. My hand, just aware of the softening into bosom, stayed still for her sake; but the nerves of the movement tingled with temptation. Only my finger moved - hardly moved in the tiniest little caress.

Suddenly a dreadful possibility exploded in my brain. That tiniest little caress was a half-conscious playing with another sort - just the most horrid sort - a lump. My hand froze. She stirred, and I kept still. Yes. That's what they say about cancer, surely? A little lump. I couldn't remember whether or not they were supposed to be a bit painful, but I figured out that they wouldn't be and that's why they get, or at one time got, neglected until too late. Her breast presumably wasn't tender. My mind reeled at this finally disconcerting thought, and I lay still and frightened - too frightened except to let her fall right off to sleep, waiting, wondering if I had imagined it.

That gentle swell from the sea changed from the almost unendurably lovely physical torment that had rocked me - it seemed now long ago - against my sexual reflection in her sleeping body. It had been pounding at my chivalry, keeping me awake so long. But now it changed to an unbearably slow torment of a different kind, distracting me, drawing out every long minute with some sort of knell.

I don't know how long it was before I could pluck up courage to move my finger, to be sure that I would not wake her, yet needing to make sure of what I'd felt in that tiny exploration.

I won't go into the deadly anxiety with which I acted out the business of ruefully dragging the time out, making everything on *Foxglove* fast and shipshape, "so that Tom would find things OK." Pretending to drag the time out, yet wishing it away. Inwardly I was in a turmoil. Showing concern about our future tangled me up with double meanings that had been so foreign to our relationship, saying one thing and meaning another.

I pretended to have some kind of ear-ache as an excuse to see the local doctor. This would seem to justify and explain the not-too-relaxed expression that I know so often crossed my face, and my desperately thinking brow.

I came away from the surgery convinced that I must tell her somehow. The damnable irony of it all bit into my soul, but tell her I had to. I tried to disguise my worry and the urgency, but, pretending to make it incidental to a (contrived) visit to Belfast, we "fitted in" a check at the hospital.

Pretence vanished when she was called back in. Twenty-four years old. Her world - our world - collapsed about her.

I sat beside her, waves of horrid thoughts showing on the cheeks where tears were half kept back. She put her hand in mine, but I knew that, to her, her arm seemed to hang from a body that was to lose its femininity, possibly life itself; with the thought whether or how that should matter any longer.

She said slowly and quietly - as I suspect many a woman has said in similar circumstances, "You'd better leave me now, Fred, if you want to. Don't think you've got to stay. You'd have to pretend . . ."

As I am sure it is with most men, I was filled with a big tender urge that comes from the deepest callings - even the deepened callings - of sexuality; perhaps from the divinest sacrificial truths of losing and finding . . . 'He that loseth his life shall find it' - and maybe, above all, from our simple human desire to pity and to rescue. Maybe it is a feeling that could be unique to that particular, rather special situation. Men will express it differently. Some will find their 'Razor's Edge', and just know that this is where they come in, and not where they go out.

However that may be, holding her hand I knew that I could have lived for this. I had thought that I had found love, and perhaps I had; but it suddenly seemed that all my life, India especially, had been a preparation for this hour. All that had happened out there was very near, it redoubled the meaning of holding those frail fingers. So I called upon India instinctively . . . .

It was there across the ocean; but an ocean, it seemed, of catastrophic thoughts. Pity, sexuality, and sacrificial truths entwined themselves with worry and fear - with trying to imagine her thoughts. I thought of that beach, that pony, that over-daring little ride, and my mind swam. An almost crucifying irony it would be if she got this, in some primitive ancestral way, to look like a punishment; as the wrath of God even. The chilling thought came that, picking amongst the ruins could be the shrewish, uncharitable village tongue: exploring; hinting at the obscene reverse of our covenant - covenanting with the Devil, maybe getting our retribution?

Suppose even, that some distant eye had seen us? Oh, my God! perhaps . . . I never knew whether some such idea of village ghoulishness, or even some inner gnawing uncertainty of her own guilt assailed her, then or after. Naked girls on horseback? Yes, I'd persuaded her, on that beach; whatever made me do that? Perhaps there had been someone near that beach after all.

But India was recent; and somewhere deep down I knew that an answering empathy and affinity were in the ruins of her shattered person. And a little bridge of hope started from there; and seemed to reach across. Her end had not quite sunk.

We kissed. She kissed me - and a quivering hope was moistened by a special tear. "You can stay, Fred, if you want to - but remember me how I was", and she looked down, and I did too, and tried very hard to smile.

A few weeks passed. We both tried very hard to smile; but there was a struggle between that and other instincts. Her struggle to find the mental freedom not to hide her mutilation from me - or should I say even from herself when I was there.

There was the perplexing, bottomless question as to whether I was perhaps concealing secret thoughts to which I would not admit. In fact I had nothing to conceal because I felt only caringly, which is completely natural to any man with a reasonable shred of kindness or decency somewhere in him. But it seemed that her groping trust on this point was half outbalanced by her dislike of seeing herself; and therefore only half effective in dissolving the worry about her appearance and freeing her from it.

I, too, tangled up my silly, over-idealistic nature. I told her - and I meant it - that I wanted to free her, set her free; that it was as if trust on this matter would intensify her femininity, take us beyond where ordinary life can hardly ever go. I knew that ordinary entrancement can run into the doldrums; I doubted whether these sacrificial truths of losing and finding - 'He that loses his life shall find it', ever would.

Then, sometimes, I could see the distress and the accumulating problems again and I would kneel beside her, and a faraway look would settle in her eyes, as she listened perhaps to a story of her being somewhere on an island, or lost in a boat at sea, and me wanting to rescue her from the sea of her thoughts. And, once again, she would smile and know for a while that her spoiled figure had by no means given any problems that I wouldn't admit to. But possibly even then her mental suffering swamped all that from view . . .

"You know", I said one evening, "this has drawn us so very much nearer together. Don't you realize that your single remaining breast, has taken on to an intensified degree all (and more piquantly) that I ever felt about this part of woman."

"Ponies", she said quietly.

We looked at one another and winced; and groped as if on the way to some stranger, profounder truth.

I never in fact knew quite what I did mean when one evening with the signs in her face all worrying, I suddenly said "Molly, let me be your breast" - except that I longed to make up for the loss.

I heard a play on the radio not long ago, entitled, I believe, 'The trains don't run here any more', in which a young country signalman had lost his leg. My eyes filled, and my memory plunged somewhere into the past, as I heard the young wife say into his despair, "John, dear - can't *I* be your leg?"

I knew that that wasn't just a play. Somebody had made a play around a memory. Perhaps they too had heard someone crying in the bathroom, and worried about them, weakened and struggling sometimes by themselves?

But - and I mean this specially - who is any man to judge? Whether a woman can sit there and watch a man and not wonder just what he's thinking; some kind of inexpressible spiritually convoluted sadism even? Perhaps just another part, a distant part, of the tragedy?

I took her to England and got her a little bedsitter a few miles from where I was living, and gradually she learned that life could go on. Something like her old self returned.

I took her down to London. I shall never forget her wide-eyed first encounter with the escalators on the Underground, and with the bewildering crowded streets. Television gets these things to everybody nowadays, but in those days so much could be new to someone from out of town. I believe I took her even to her first cinema.

She snuggled up, and I was happier for her; quietly intoxicated with watching her half-childish excitement at all the sights and sounds of the big city, and feeling that it was all helping her youth to win through again. I soon sensed that London put her on her mettle; that it was not only an excitement but that it had taken shape in her mind as a challenge.

My old days of throwing my dirty shirts into the bath-water and walking up and down on them, and of boiling an egg in the porridge, or trying to toast it, were over. I had to give up peeing on the landlady's aspidistras: a plant for which, as a symbol of bedsitter gloom, I had developed an obsessive hatred. I got over that.

A more deeply ingrained affair is that about every tenth night, about twenty minutes after I have got into bed, I begin to feel my toe-nails growing: and the pressure gradually mounts up until in desperation out I have to get, and find the scissors. I might add that this is one way of drawing upon one's partner's very last reserves of patience, stretching out their nerves and screwing them up so that they are left grinding their teeth as the hundredth sheep approaches a jangled stile . . . So what do you do? Plead that this must be something to do with the changing humidity down at the foot of the bed? That scientifically speaking, it isn't the growing I feel - that they just dry out after having warmed up, and get curled solid the wrong way? It would be a sad world if things didn't grow in bed.

But there couldn't have been a hassle more redeemed by tenderness than the trouble I had getting Molly used to the new shower: the unspoken labyrinth of hesitations, the unspoken understanding. I guessed her real thoughts, the untried contrivance, and about the curtain - but how else, than by a gentle humour, could healing have come to her - finding the soap, then getting it in her eye. But how else could I have got such a mentally quivering smile, then such reassurance from a kiss?

The following winter was severe; everything iced up, and everything rationed. The level of this rationing is hard now to visualize: tiny pats of butter; forms to fill up to buy clothing, bedding, etc., and all sorts of complications that I can't remember. Some Friday afternoons I used to have a cup of tea in all three of the local tea-shops in order to smuggle three tiny pats of butter, the size of a modern 2p piece, to my mother for her weekend.

The Two Himalayas

One read the papers about Europe; about cycling twenty miles to get 5lbs of potatoes; refugees from the Russian Zone flocking westwards, without warmth or food. There were tales of clever operators in the Occupation Forces in Germany buying flats - some said blocks of flats - for blackmarket currencies such as cigarette-cards. I've no idea how much of that was true.

There was a lot of heart-searching about Hiroshima, the Burma Railway and Belsen; and descriptions of our fire-raids that sucked people caught in the draught along the streets to be incinerated in early medieval towns, like Dresden, with its ancient wooden houses.

But to me the only people who knew where they were in all this were the Quakers. Half intentionally and half by chance, Molly and I found ourselves watching with admiration as they plunged almost calmly into the turmoil, joining up again with their opposite numbers in Berlin, and other inconceivabilities, accepting and confiding in one another after it all. Nothing seemed to have broken their mutual trust and the allegiance of their fine, clear minds to those in need, to one another, and to their vivid sense of each individual's right of access to God.

I remember our sitting over a coffee in a strange house in North London. The phone rang. Yes, Rosemarie would be arriving on Wednesday. Who had better take her? And they had a little discussion. Rosemarie? I asked. Yes, she was coming from East Berlin. Admittedly this was before the era of the Berlin Wall, but the mere idea of it set my mind agog. That sort of thing just didn't happen in those days. It shattered all one's usual parameters.

I tend to get over-simplified ideas about history, but I thought I knew enough to have a dread of another 1919 Versailles-type Treaty, where the French view had prevailed, albeit understandably, against the more magnanimous British and Americans. As I saw it, this had sown many of the seeds of the 1939 War; and it may well be true, if a little simplistic, that Hitler fed upon the fact that nobody had much faith in the Weimar Republic because of unemployment, lost colonies, reparations and so on. Perhaps I was soft - but if that was the case, then men like Woodrow Wilson and Churchill were not wise or statesmanlike: they were soft too!

A Quaker who had been over to a teacher training college in West Germany came back with the rather stunning conviction that many of the young trainee-teachers there would almost have given their right hands to come and visit this country. Not for the holiday and the odd slice of bacon but because they saw us (then) as the only hope for a world that had broken to pieces round their heads; that stood for something amidst their disillusion. Their lost Fatherland needed them still, though. The time was to come of course, and before not very long, when these same youngsters began to feel that we had missed our opportunity. Maybe that is half-true. But at the time it was an intriguing thought, temptingly evocative of a decent sort of pride in this country, debt-torn though she may have been. What was the alternative, anyway? - the old 'guns or butter' - guns, revenge; more killing, then more revenge - instead of healthy young kids?

I read in *Picture Post* that the children in Germany went barefoot, sometimes, in sub-zero temperatures, and on about a third of our rations - if that. God knows, I was bloody cold myself. Molly and I looked at one another, there in the North London house. This was too much. Perhaps I'd said that before about a poor little bastard in a hotter country; but here we were together, and it was too much.

The Two Himalayas

With the help of the Quakers we collected a few little shoes from the dusty top shelves of one or two local shoe-repair shops, those nobody had called back for, bundled them into a sack, and posted them to the teachers' training college for their barefoot local kids. It was the willingly-acted-upon hunch of a quiet moment.

Here was the Lakshiman feeling - getting off that train - all over again. But shared with her. To our amazement, other little repair shops were to just brighten at the idea, obviously so warmly pleased to do it - one even reached up and gave us ten new pairs in ten unopened boxes!

That night was very special. She came towards me quite forgetting her wound and the loneliness of her single breast, because the day had made her happy. And so, my God, was I.

A little while passed, immersed in this more restful happiness. I think she was no longer living in quite such a hidden tangle, badgered by quite such thoughts as being a self-condemned potential outcast, accepting as her fate that no man would ever see her again - or want to see her - or that she could ever face being seen, her whole life . . . suspecting that men conceal the true nature of their thoughts.

She could hardly know it, but men whose loved ones are so afflicted mostly find that they are moved to become some kind of saint, just for once in their lives.

After much heartsearching, I arrived one day with a book, very cagily chosen, on massage. It had an Indian flavour and it could open, perhaps in the most honest way, the door to a dream - the final release to physical floods of true forgetting, to a little Tristan even. It is a healing thing, with which few are honoured. Hopefully, I romanticized its possible hidden secrets, and grafted what they might have been on to those ideals of trust, perhaps sliding in and out of sexuality.

That book was remarkable for being both therapeutic and sensual; equally remarkable for that which it avoided, as for its obvious assumption that the reader would have such mental poise that a towel would not be necessary. I learnt for instance that each toe was to be alternately pulled, then pulled and twisted to right and left - but not out of its normal line; tickled very gently down each side (a strange and lovely feeling where one would hardly have thought there were any nerves), and then bent over one of your fingers. This feels funny, which gives relaxation from possibly 'heavy' attitudes to starting into a subject like this.

I remember that one is supposed to ease the blood in the limbs towards the heart, to knead them very gently, to lift a limb with both hands; to kiss an elbow (inside), to co-ordinate your two hands, fingers or palms, in circular motions that can please even the tummy, and several other two-handed ways to making the patient's flesh feel warm and alive. A very short chapter on erotic massage was confined, strangely but intriguingly (and genuinely) to the erotic intricacies of the face and head, navel and knee and spine - tickling the inside of the thighs not being given this distinction (which I found pretty 'cool', betokening a poise somewhere between a way-out desperate build-up, just a way-out poise, or a way-out trust). She, Molly, had a way-out cheek. With something of that old twinkle, I got "May as well - instead of the ironing". And, puckering up her nose, "Just as boring . . ." - from a girl who still went to confession, but who would reach out for that little bottle of oil, too; usually baby oil.

The Two Himalayas

It was a very difficult thing to forget this genuine innocence. There are no words. You can't believe it; it feels as if your inhibitions are somehow slippery; life-long fixed inhibitions sliding away . . . Perhaps because the oil was warm?

But there's something spiritual here; and she'd go quiet.

This strange, almost paralysing, almost dream-preoccupation state would stay with me, only just allowing stray bits of ordinary preoccupations to enter my mind. Even walking sometimes along a busy road . . . what was that going on ahead?

Suddenly I was jarred into the present.

I shall never forget that lorry-driver's face. "The little boy's alright", he kept saying. "The little boy's alright - he ran under my front wheel. Didn't nobody see it?" he asked imploringly. He was white and shaking as the crowd I had joined grew - round the lorry that had jack-knifed across the slippery wet road, and bashed a lamp post down.

I didn't want to keep Molly waiting outside the tube station. It was a bit windy, and I had another minute's walk or so to get there. And, as the police and then an ambulance arrived, I felt as useless as the rest of the crowd, except that I didn't want to get in the way. Perhaps on a quiet country road you could even help in some way; but here it would be blankets and a stretcher, and people craning their necks to see . . .

A policeman was getting into a car as I started to walk away. I was stopped in my tracks. A wave of intense unbelief froze me as I pushed the thought from me. I knew that greenish little bag so well. Another policeman came over to his car with the sad little remnants, the unbreakables; the bent up toilet roll seemed to tell of some terrible intimate injury. God! What had happened? But as the question rose and took shape in my mind it was answered, nearly, by some voice in the crowd: "I expect she's dead".

"Not quite", said a kind-looking policeman who took in my situation and overheard the remark at the same time. He hesitated. "Are you . . . ?"

"Yes", I said, "that's her bag".

I have a dim recollection of a ride in a police car to the hospital; a vivid one of waiting - it seemed endlessly; and occasionally answering some question; of the sense of foreboding that was soon confirmed by the sign 'Intensive Care' over the doorway.

"She's had a mastectomy", I said, rather helplessly, as if they didn't already know; and as if it mattered any longer.

I thought of her struggles about being seen, about no man ever wanting to see her ever again, and the irony of that thought, as she lay there, never to live again, became mixed up with the past irony of discovering that lump on our very first night; and all this as our new happiness and mental freedom was just beginning.

It was all so painfully cruel that I knelt by her bedside and my world collapsed, and my heart broke. I suppose that some poor embarrassed nurse showed me out to where men are usually given cups of tea before they carry themselves off, I suspect more breavely, to an empty home.

Quite a long time passed. I can't remember much about it. I suppose it was full of meaningless gaps. Moments of blank staring at the empty room; the crowded but uncaring streets.

The Two Himalayas

I wondered what had happened to those odd streaks of orthodox assumptions about bereavement that we think we believe in. They would come and go.

But gradually I began to feel that she, Molly, had somehow left something in my care - this strange mixture from the Himalayan gateway and the Irish creek. The knowledge that the ego and, in particular, undue concern about one's reputation are stumbling-blocks, wasting life.

The strange revelation about the power of willingness, that had occurred at such an intimate moment on the boat. That this was the key to opening life out, not being spiritually pushed around.

I was aware that little seeds of boat-hallowed thought started to ripen into nice little happenings, that slowly persuaded me out of the earlier pit of irony. Experiences I had never had, and could never have had, before becoming caught with Drummond ideas.

6

Once I found by the window a few packets of crisps. They were a bit special because they were all that was left from a so-carefully wrapped present, carefully wrapped by a real and very attractive princess, who, intriguingly, didn't take exception to sleeping on the floor. The little note with the present said 'lost of love' - a touching effort at 'lots of love'. And to my friend, 'your picture is look young ...' We still keep in touch. Funnily enough, I still remember the sudden inner rearrangement of priorities that led me to turn round that day, to go back and call at the Save the Children refugee centre down that lane. Funny, that's how it seems to work; in odd ways. Turn your priorities round; turn the car round; turn the future round.

But it would all sometimes dredge back down to remembering those special chords of my younger longings, that sort of young holy dreaming Grail of the shy beginnings of love. Dreaming of melting together in a mutual secret of no taboos. Saddened by my memory of our having lived through this together as in a kind of dawn, reflecting each other's shyness ...

But eventually, deep down, as with that old story of Excalibur, it was as if a hand had appeared above the recent lake of bitterness, holding the strange truth - strange, yet profoundly clear - that our discoveries under that candle could even survive the unhappy requiem and end of our pilgrimage. As if the emptying of that Holy Grail had left a candle burning that is greater than ourselves, whose meaning is for the whole guidance of our life.

I thought of that boat scene; that faded little book again, lending something to me; looking down from the unknown to that wildly tempting moment that I had never known, to give a serenity that I knew was of God.

Somewhere, too, was the dawn of the idea that there is something in this willingness that captures an elusive serenity that inner man has hungered for down the ages. Try it. Try that which doesn't necessarily make sense. You'll soon find out where the little bug of contentment lies; the 'of God' contentment.

Anyway, I tried to see what it meant for me. Ungrudging willingness? Some Samaritan mission perhaps? Or, more paradoxically, a passport to breaking away from one's present mould, from any present mould.

In practice, I got to poking about with my daily life amongst its hunches, quixotic or boring, until something - whose justification was probably rather wispy - popped up which had a nebulous 'of God' ill-defined attraction - something that wouldn't quite go away. I realized that that was for me; that was mine.

Philosophers must of course have their proof and pester the thing with questions.

But it will explain itself in its own soul-kind language in which conviction comes home, somehow, in a different way.

We must have our logic. Yet I have since come to see that I had in fact found my path to an old truth that down the centuries has mystified all who have sought and not found. If they wriggle away from the apparent indignity of this inner willingness, then they can scarcely find any rainbow's end. *It* will wriggle away from *them*. Or they are preoccupied by the ramifications of their egos, and what sort of figure they cut, or don't cut. (Whereas Francisan friars, for instance, *know* that they have found something in their way of life).

I wonder. It takes me back to reading that book, sitting on that gate . . .

I found myself continually feeling back to those memorable first days in England after the war, before sailing to Ireland, and how that old book had devastated me. I didn't know then that it was to be somehow consummated by that kneeling covenant on the boat.

Those first days of settling back into this country haunted my mind. Life, a busy life, and the tired mind that sought refuge by our River Mole, always throwing those new attitudes - those Drummond attitudes - towards the stars, and travelling through a vault of answered questions.

For a little while, almost inexorably, as time passed and time healed, I tried to live out what those Himalayas and creeks had come - would come - to mean to me; except that I didn't *try*, it just was somehow in the air.

Then was to begin a series of things that endowed religion - this kind of religion - with the atmosphere of a natural and convincing law, yet left one not being too pious; feeling instead rather spiritually comfortable, spiritually cosy.

As if to puzzle me further, and at the same time add to my wondering conviction, I found myself meeting up with odd, unusual people; interesting, different. A princess; an Indian gentleman who was into Dr Schumacher's intermediate technology of 'small is beautiful' - whereby the simpler Western inventions replace rusting complicated tractors that no-one knows how to fix, and puts unused manpower back on to the land, and into brickmaking, pumps, and hope. This Indian gentleman was somehow ultra-fulfilled, having uniquely dumped the pride and mystique of westernization in favour of the villages, even though a large 'BSc (Failed)' over his shop would have made him more money.

There were others; and before this rather crazy Franciscan bug had been smothered, I was to see the inside of Number 11, Downing Street, where I had a chat about what youth was really groping for, and got into the beginning of the Christian Aid movement.

And later, I heard in the office of the Bank of England's Foreign Exchange Department (huge, panelled and overpowering) that he, the boss, couldn't bend the Exchange Control rules: although it was only about a little money collected to bring two or three of those young teachers from Germany. And his daughter lived just down our road.

Regretfully I'd now have to try and uncollect it . . . but pehaps the Foreign Office could help . . .

But *who*, in that place, would . . . ?

The Two Himalayas

I remember getting a refill of inspiration from sitting for a while in the empty local church - seeing the issue in perspective. I came out into the sunlight. Quietly certain now that the Foreign Office must have some susceptible human beings behind its massive walls, I took a taxi. Walking along the bewildering corridors I spotted a likely-looking department labelled on one of the heavy old polished doors. I knocked. They didn't seem too used to odd visitors, but a youngish man invited me over to his desk.

When I had finished explaining my dilemma of the unusable, and unreturnable, money to him, he said something about my name ringing a bell. Was I, perhaps, the chap who'd written that letter some weeks ago in *Picture Post*, about sending shoes to the frozen German children?

A little amazed, I said that I was.

He wrote a note or directive to the Bank of England, got it signed, and back I went.

And so it was arranged, and the three young teachers came . . . In fact, they were pounced upon, and bewildered by kindness.

I've gone into this in some detail, because it stays in my mind as a particularly impressive example of the strange 'lucky breaks' that kept happening. Lucky? Luck composed of responses from people, sometimes only marginally involved people; sometimes from people you had no contact with, or at least did not know at all.

If a period came when motivations got confused and my little times of quiet susceptibility tailed off, then these things - these responses - just stopped happening, or else were lost on me. I'll take quite a lot of convincing that it all was luck, that there wasn't now and again a suggestion of unaccountably *too* much coincidence for it not to have some other significance.

The sceptical can have it that it could have been some kind of subtle personality change, which I suppose could change your *luck* in odd ways. But I believe that in sometimes bucking the system of selfishness, one seems to get into a world that is otherwise closed.

Perhaps Ulli - wandering years later in the hills, and whose spiritual hormones were 'Love and Peace' - found something fascinating about spiritual valleys; valleys of 'unegoselfishness', and their secret laws and inspirations.

Maybe, somewhere on the hillside overlooking that old river, the Alph of our kitchen poem memories, there ought to be a kind of memorial to Drummond. I can't quite explain; but it may become clearer.

I wonder, in passing, whether the Nazarean Wanderer could have sustained His life by some kind of spiritual conjuring trick and found that inner call for hungry Man just the same? or that unplumbable conviction of unplumbable spiritual ultimates, which few would deny were his.

I think perhaps He lived out *laws*. For instance, that the reverse of pride is linked with the reverse of agitation.

One friend of mine, who'd done a lot of hitch-hiking, had an idea that I acted upon. He said go and pick people up (petrol being short) and take each one of them where they want to go, and see where you end up.

I stayed two nights at the Franciscan friary at Cerne Abbas in Dorset. A 'pick-up' was on his way there, so that's where we went. I remember it all clearly: the Harrow boys playing table-tennis with the Borstal boys; the friars' scintillating conversation and sense of humour after the day's silent periods; the evening meal when a book was read to us; the little cells; and the friars who lived partly there and partly with the destitute of Hamburg or London, or who tramped the roads of England.

I remember the sense of peace and community, each brother having some kind of trade to keep the place half-independent from the outside world. I had the impression that, while they accepted the evil of the world, they went about quietly accusing people of being better than they thought they were, reading something in their hearts rather than their lips - seeing beyond the petty egotism that is their 'front' to the deep desire most of us have to love and to care. Friars are of course a shade less cynical than some but they stuck in my memory.

Continuing on my way, if that is the correct expression for such a sort of convoluted taxi-ride, I then found myself invited to what seemed almost to be a castle - there was a moat all round. Then I was taken, somehow, to a cosy little thatched cottage on the banks of a river where we threw away the papers and stopped the clock, and I was so peaceful that I learned about the agrimony growing by the water, and lots about the history of the old ruins there.

It was a holiday so absolutely different that it comes straight back to my memory from all that *post-war* time ago.

More time passed.

As life went on it broke up the log-jam of those old Irish memories. Perhaps it gave them a less hurtful meaning, because one by one they moved aside to allow the feeling of freedom and desire to come through, so that I left another lump of my heart in Germany.

I remember volunteering to go out to Pakistan to help with some disaster, leaving my job; but they wouldn't have me.

I changed my digs instead; but my motivations ached of Drummond, tangled with Germany.

Tangled with Germany? Germany 1947?

I'd heard somewhere about a Fellowship of Reconciliation Movement (Quaker, probably) who were holding a conference not too far away.

Three senior students from Oldenburg Teachers' Training College had been invited (they didn't want streams of future students being embittered by embittered teachers.) This conference was to lead to our marriage, a year later, Karoline and I. It hurts the acounts department of my brain to recall that I pawned my watch to get there, just across London! But that's how it was. And when I'd caught her dancing by herself in the moonlight with, fascinatingly, the tiny myriad blossoms of some common but beautiful hedgeside weed in her hair . . . well, I suppose I thought it all uncommonly attractive.

Attractive, too, was her fascinatingly colourful life. Serving sometimes in the trenches near the Russian Front, she had lain her sick mother on an old door and, with the help of her younger sister, had carried her through the woods in the night out of the Russian zone. Her father (a surgeon-commander in the German Navy) had become a prisoner of the British near Kiel. Karoline, hidden in the coals of the engine-tender, crossed North Germany to visit him, and was arrested while climbing the fence of one of our POW camps to hand over a few scrounged apples for him.

The Two Himalayas

Later in Surrey, in quieter times, we two scrounged a cup of tea one Sunday afternoon from that same camp commandant at his peaceful little countryside retirement home. I hope he saw the funny side. I did - especially as I'd just had to ask some high-ranking army officer for permission to marry in his Sector; and I'd hated it. Ask the Army if you can marry? A bit difficult to swallow. Except that the ceremony was conducted by a kind of blackmarket priest.

Incidentally, it has always fascinated me that Karoline's father had been the gunnery midshipman, range-finding up in the masthead of the Imperial German Navy's flagship in the Battle of Jutland, during the First World War, while it was the young Lord Louis Mountbatten who did the same in the Royal Navy's flagship.

So it was that the first few years of our married life were spun on a thread of the International Reconciliation Movement - nowadays rather old hat, but then pretty well unexplored territory.

It took the weight of our two pasts, as mutual love took the weight of our mortgaged half-open door.

And, meanwhile, Ulli was born, to become another memory: the most pressing reason for this book.

He was a child of a time when the world was licking its wounds. He didn't really know that the aunties who came to see him came via Harwich-Hook, and Heaven knows how many young people from there (where the lights were all out) wanted to come over here and think.

It is the accepted thing nowadays, and indeed works in both directions, but then it was a break-out from their world of near despair to that special part of the world - this country - where they saw hope for the future enshrined in our way of life (a way, you hoped, still not too crumpled up).

I have written this before, I know; but it is still startling how they were drawn to the world of recent enemies. Some kind of respect had filtered through it all. Strange things had happened. A bloke in Wilhelmshaven told me about the Royal Navy's last signal in 1919, as his German squadron lowered its flag: '*Auf Wiedersehen in besseren Zeiten* [See you again in better times]'. Or the service of full honours accorded by HMS *Devonshire* to the *Bismarck*'s crew, wrapped in their German colours. And some of our lads were seen to be crying.

And then, hardly noticeable in a German post office, the young man who had come up to me and, bewilderingly, insisted on shaking my hand, saying - with quite spontaneous sincerity - "British Navy, I like to meet you." Somebody's friend, I supposed; but that wasn't the case: it was because *I* had a beard and *he* had spent three years at sea in submarines!

Yet it was a challenge for those youngsters coming here then, with uneasy overtones; and an act of faith in man's forgiveness for man.

A year or two after the three young teachers came over here, by grace of the Foreign Office bloke, that same college in Germany was asked if any more would like to come. Twenty-one students said yes.

The only interesting point, after all this time and a million visits, is that this was by grace, I think, of the power of odd-seeming little hunches that aren't fluffed out and smothered at birth. So, it is worth considering the value of nestling into an unconventional attitude - spellbound a little by watching the evaporation of one's ego - and letting a different motivation take over.

Willingness: the magic word - my favourite word - that unconstipates the drudgery-bound soul.

We lived in a seventeen and a half foot long caravan - the same length as *Annabelle*. The official said that we had to take responsibility for the board and keep of the visitors, for the whole five weeks they were to be in this country. Yes, I'd do that (thinking desperately which of our friends might have got some sort of tent we could borrow). Twenty-one of them. I signed the form. I suppose, officially, they all lived for five weeks in that seventeen foot six inches.

The same strange flush of fulfilment came over me as with young Lakshiman, and then with the sack of shoes. I was very happy - we were very happy - and very peaceful, with that special kind of knowledge that haunts and belongs to fulfillment. Not, I believe, the same thing as indulging oneself as a 'do-gooder', which can go right and yet go wrong inside, but the peace which comes to those who have embraced something like 'willingness', giving that strange obedient feeling an authority over all other feelings, and who answer *that* helm.

It all seems a very long time ago now, but I believe that it was a pilot experiment as a goodwill venture between the two ex-warring nations. We wrote to lots of friends all over the country, so that by the time the group had arrived, each pair of youngsters had two or three different homes spread all over the country to which they hitch-hiked and youth hostelled.

I say that we wrote to lots of friends. Once so rare in my life, friends - and people with whom I felt a strong common link - were a phenomenon associated with this new lease of life, a development I could never have dreamed of. Strange. All really through giving best to the murmerings of unsettling ideas; through making oneself open to the rustlings of ideas that you have to wipe the smothering, flowing self away from. But ideas, unborn promptings, that may be the 'Hound of Heaven', may be *you*, may be of God. Which I suppose, in that quiet moment, could all feel the same.

We started looking for tents for their fortnight's stay, but no-one slept in a tent. Every one of them was taken into a home. The local congregational church folk insisted.

It was moving in the extreme to hear the Lord's Prayer in German, the youngsters filling the first two rows; then a German hymn. No one nowadays could really grasp the effect of that. Many hearts were full of feelings; eyes were moist at the reverberating echoes between this and the recent war. The place went still: to store everything, I suppose, in our memories.

And so, almost inevitably, another group grew up, and arrived the next year; but the Quakers, the *real* pioneers, still carry on where there is need. I imagine that they weren't far away when Ecumenical Centres were thought up and started - meaning centres that span all denominations, both in this country and in Germany.

One of my memories is of a young teenaged girl named Hildegarde, who was staying at one such centre near Hindhead, which seemed to collect those who had a kind of 'Scarlet Pimpernel' background to their lives.

The Two Himalayas

Hildegarde's mother was a Quaker living in the part of Germany that was overrun by the Russians. She sent Hildegarde away, but had refused to join the exodus herself because of her deep pacifist convictions, which while encompassing passive resistance, were motivated by 'Love thine enemy'.

When the Russian soldiery approached her house, she had sat there playing Beethoven on the piano, all alone. About a dozen Russian soldiers had gathered around her, very quiet and very moved; and these few coarse men - whose normal behaviour would have been so frighteningly crude in the ways of the primitive Russian of that time - those men protected her for days from their fellows, and later escorted her to a place where she would be looked after . . .

I felt it a privilege even to wash up with the daughter of a woman like that!

'Scarlet Pimpernel' type adventure, seems to centre around people with deeply held convictions, who don't treat barriers as impenetrable, and who get other people across them, and care for them at astonishing places like the Hindhead Centre.

After being offered a job as joint wardens of the first such Centre in Germany, and it all coming to grief over those same Exchange Controls that had so nearly scotched the student invitations, our life seemed to settle down around our own little problems. So that, from this story's point of view, there was less and less in the next few years of our married life worth recounting, in that it was all about how to cope with a young family on very little money, which is what most people seem to be trying to do.

Having a child to look after in a tiny caravan on the top of a hill is difficult, so we bought the next-door gypsy caravan. That's where we used to put young Ulli to sleep, and cart him back in with us later, when it started to get dark.

We later got hold of a wild piece of land, an impenetrable jungle of blackberries that sloped up at some ridiculous angle, and managed to get a house built. We picked blackberries from the scaffolding, and stewed them in the caravan parked among the other blackberries.

I built a bridge from the top of the house to the top of the slope. That didn't collapse, but my health did. After the third (hitherto undiscovered) type of dysentery was cured, things got better.

I got a new job, and little Janet arrived.

Time passed.

There was not much 'let up'; nothing quite like that "agrimony by the water".

I used, over the years, to go down to see Tom, and that beautiful garden - still "ridiculously beautiful" from the way it had first overwhelmed me. Gardens don't usually do that; but this lovely place was set on a steep hillside, where three streams joined together and divided mysteriously into two under a little bridge, with two half-tropical ponds, and mossy banks between the tumbling little waterfalls. One could lie naked on the moss next to where one of these waterfalls found its way into a rocky pool, in which one could just stand surrounded by the clear water and the rocky sides. Although I went there many times, I could never be quite sure how to find my way through all that near-jungle of just-separated, towering, exotic flowering bushes and trees: that monument to the genius who had found them and planted them some fifty years before.

The Two Himalayas

The old gardener seemed to be a living part of it all, outlasting even the unevenly-roasted girl-friend. He used to walk around with an ancient trilby hat, followed by a cloud of flies keeping station just above him. Surprisingly, perhaps, he had accepted the occasional friends who sunbathed nude on the lawn or by the pools. But the two or three girls who used to go there had a horror of his taking off that trilby somewhere in their vicinity, lest the flies become disorientated - and settled down and tickled, I suppose.

Sadly, over the years, Tom had begun to look pale and drawn. He and I would sit up till the whole countryside was long since in bed, trying to figure out his chances of financial survival - among the incredible ramifications of Capital Gains Tax, Capital Transfer Tax, Forestry Allowances, Leasehold Enfranchisements, and Heaven knows what other traps there are for the rich man who backs a struggling venture. Devices that threaten, or threatened, the dismemberment, or the sale to some over-fat merchant, of every specially beautiful heritage from the past.

I would scrape up the fare, and we would sit over a tiny fire, blankets round our shoulders, wondering how long he could go on living in that place he loved, and which we all loved. He never mentioned Molly; nor the boat, and I knew he must have sold it.

Not surprisingly, his face grew haggard. His tough constitution was crippled by a stroke; and then another.

And so he died, stuck up in a bedroom: and the garden gradually withered, and it all got sold up.

My second home.

I wonder if I shall ever park somewhere along the lane, walk down that bend; maybe even trespass? Find the ponds, and try and sense if I could knock on those old oak doors, one inside the other; or sit in the church-like porch that embodied our two such separate natures, Tom and I.

Time went on, and with it grew a sense of having somehow lost my way and my convictions; the not quite useless passing of the years.

There was something I wanted to find my way back to, to share; to rescue from getting lost for ever in life's silly passing complications.

So once I went away for three or four weeks, staying alone in a little wooden chalet amongst the woods, which were part of a naturist settlement. Here there was a natural lake for swimming, and you could lie alone in the little clearings of blue anemones amongst the silver birches.

I had been there before, in an earlier period of restlessness, and I know that there's nothing like that for relaxed thinking. Perhaps the freshness and stimulation of nakedness in the woods helps cut away the trappings of life, the distractions, the newspapers, the television, the clogging effects of civilization. Maybe even there's a sexual stimulation somewhere in this that makes you feel more alive, certainly more in tune with nature.

One or two of the things I did might perhaps have been cranky, superficially so, at least. To me, though, they had been rather more odd expressions of wanting to become more alive, more aware of inner convictions seeking a freshened mind.

I remember waking one morning at dawn and feeling it a strange challenge to lie naked and quite motionless on the soft turfy grass in a little forest glade, arms stretched out, for the special hour before the sun came through. It was terribly difficult because of the little creeping and flying things, and the vividly-imagined wormy things. But it was an hour crowded with intense life, virginally one with earth, mind over waves of convention-ridden thoughts; experiences lost to mankind but surely fundamentally belonging to man. It was a sort of baptism into nature and a tiny yogic christening at the same time, which I always look back to.

One late evening, too, that same old primal challenge came back, and I found myself gradually becoming fascinated by the idea of diving off the thirteen-foot high diving-board into the lake in the dark. My previous maximum - and in daylight - was about three feet! I suppose it was some kind of symbolic attempt to prove that my mood of spiritual adventuring could overcome even a lifelong fear of diving. At the time, I had been to elementary yoga classes (to try to get fit, cure tummy-aches and, secretly, to womanize). Somebody there had spoken of how a relaxed body can still the mind. Perhaps I dared to imagine that a stilled mind could relax the body's fears? They'd even talked about 'levitation', whereby in some way-out kind of relaxation your body feels as if it is floating up to the ceiling. Perhaps I was going to try levitation in reverse? Just feeling as if you are dropping. The uncanny world of a stilled mind.

This is no doubt poor stuff compared with the heroics of real adventures; but to me it was quite shattering and unbelievable that I was to feel my way in utter darkness through the woods to the pool, impelled by this strange impulse. I removed my clothes, climbed the invisible ladder and edged my bare feet bit by bit along the plank until one toe snuggled gingerly over the edge. Lost in the dark nothingness of no more plank, I suddenly embraced the empty air, the plank left my feet, and I went down through darkness to the invisible water.

Nobody could have been more surprised than I was. I had to try it again; but this time I landed painfully on my sprawled-out front, decided enough was enough and made for my bed - leaving nearby hut-dwellers talking about a sudden splash and "could anybody have got drownded?"

And then there is my discovery about the bubbles. I think even naturists, who don't need convincing about the liquid charm of swimming without a costume, could be surprised by the elegance of this futher contribution to the matter.

I refer to the fact that if you stand naked but quite still in a pool with a muddy bottom and then very gently stir one foot, nothing could be more exquisite to the tickle-nerves amongst the tiny hairs of your whole body than the gentle lodging and gently escaping little bubbles. Try it.

The keynote to the whole experience seemed to be 'Let not your heart be troubled'. Somewhere behind it all was still the feel of the harmony that Drummond hit me with, and that had "fucked Ulli's mind" - as if he had climbed a mountain of thought, and couldn't express how much it had all impressed, and perhaps shaken him. Somewhere behind a palpable unstress was 'Cause and Effect', that gave a kind of overflowing assurance that you'd still got one foot on the ground.

I lay near the hut in the sun, seeking back to those memories. You cannot forget. It's not quite the same thing, but I do feel that there's a striking congruence between the vague chemistry of those convictions and the words of Albert Einstein, the inscrutable and daring thinker who turned Man's concepts of the universe on their head.

The Two Himalayas

He wrote, 'I, at any rate, am convinced that *He* is not playing with dice'. And he described his religious feeling as one of 'rapturous amazement at the harmony of natural law, which reveals an intelligence of such superiority that, compared with it, all the systematic thinking and acting of human beings is an utterly insignificant reflection.'

Insignificant too, perhaps, are our exquisite doubts - mostly about our more exquisite senses.

I had more than once, over the years, wandered off alone to the coast of Essex, where the lonely stretches of sea-wall and the wild tidal saltings seem to mean a special something; to restore my dulled and fed-up mind, knowing that thoughts rusting away, through not being lived by, could come back into their own.

In particular, I recall one evening on the sea-wall not far from St Peters' little chapel at Bradwell-on-Sea. This chapel was one of the two earliest landing-places of the first missionaries who came to this country. It still stands very much as it did in Bede's day, I believe about AD 650.

This part is usually very deserted. The wide saltings - saltmarsh flooded by the tide - are part of a bird sanctuary, and it has a strange contrasting quality given it by Bradwell's nuclear power station, just discernible in the far distance - so that the old values enshrined in the ancient chapel seem to stand more time-spanning than ever, silhouetted and quiet on the sea-wall. You get the uncanny feeling that they will outlast man's frenetic burrowings and buildings, and pathetic rushing energies.

It had been a warm afternoon and I had been sunbathing in the long grass on the landward side of the sea-wall. As evening came on I felt compelled to continue lying there and watch the day fade out. The chapel had nearly lost itself in the dark, while the glow from the distant power station gradually took over from the dying sunset.

Never before had I lain naked under the stars, I thought. This I must do, even if only this once. A cool breeze arose, but to my very pleasant surprise the long grass and the sun-warmed earth kept me snug and warm; a phenomenon that I had never expected. Just because, I suppose, we know so little of the nearness of nature.

As the stars brightened it seemed that one penetrating thought led on to another, once again that woven structure of conviction that runs through this book. The almost uncanny intoxication of that harmony which you stare into if you trust the concept of natural law in the spiritual working-out of life. That had got into me; into the dulled brain that had driven out across Essex.

I somehow had - not quite to celebrate - but to *do* something; something to symbolize the great sense of being alive after having come here not alive; of being awakened to a sense of destiny in what St Peters' stood for, that its reinterpretation was as vital, proud, and relevant a thing as was the power station that captured the distance.

So, being at that time an amateurish practiser of daily yoga-type stretchings and bendings, what must I do but dare the night air and do a bit of this yoga on the top of the grassy sea-wall. It sounds stupid; but there wasn't a soul for miles and it was really fantastic. And when I got to the bit where you stand on your head (and shoulders too, in my non-athletic case), and I looked up and saw my pale body against the stars - and knew that I had broken through all the bloody silly barriers that man makes for himself - the cool wind and the feeling of mental exhilaration sent my spirit thrilling into the night, from value to value. I think I was somehow aware of a tryst with that strange altar, the salt-marsh, where the real Me still meets the Tide.

I found my clothes, but I left my keys there and had to go back for them, which was not quite a spoiling but a sinking in of it all again. Luckily I found them. The stars were bright and showed me my nest in the grass. Again I passed the little Chapel, and went inside in the dark; said 'Thank you'; found my car; and blinked at the dim lights in the snug little bar of the 'Queen's Head'.

And yet I remember lying there and wondering about those other words from Einstein that, perhaps surprisingly, had drifted from somewhere in the past back into my mind; 'If I were out in the stars and galaxies, I wouldn't say anything; but the feeling would be prayer'.

Stretched and alive as it were between these simplicities and infinities, I had had to do *something*.

A pint of beer connected me to the human poetry of the pub of my erstwhile jaded eyes. I sat quietly and ruminated on the dark old wooden seat of the bar. The food was good. I gazed around me, and had a coffee, and talked with the power station engineers - talked 'shop' with them. But even that didn't make it all a dream.

It wasn't all a dream. Einstein didn't make Drummond unreal, the power station's prerogative of background presence didn't shatter those sea-wall thoughts.

Einstein was a modest man. Drummond had something of the daring and the simplicity-seeking elegance of mind of that great thinker: the unhappy discoverer of this nuclear power, the man whose theories about the universe unrolled new conceptions that are still being proved; space telling matter how to move, and matter telling space how to curve; gravitation bending light and so on. Utterly baffling! but leaving the scientific world bowed down and lifted up at the same time. So that both Einstein's and Drummond's thoughts lead one to feel, with Wordsworth:

> A presence that disturbs me with the joy
> Of elevated thoughts; a sense sublime
> Of something far more deeply interfused,
> Whose dwelling is the light of setting suns,
> And the round ocean and the living air,
> And the blue sky, and in the mind of man:
> A motion and a spirit, that impels
> All thinking things, all objects of all thought,
> And rolls through all things.

An impossible poetry in those troubled areas of man's inner being, discovering spiritual laws echoing something deep and vital in Nature.

This all seems very far from that kitchen long ago, but haven't we all found ourselves going silent while staring at the fire . . . ?

I suppose I'm trying to say that these attitudes and beliefs - the outcome of those 'watersheds' of life that are enshrined in my memory - are not just disembodied theories, isolated propositions groping around in an uncertain vacuum. They integrate and interfuse, I think, with what in the past has moved the poets of the human spirit.

They are ultimately convincing. 'Law' taking on spiritual overtones, 'cause and effect' invading us at a deeper level.

Sadly, we seem to live as if these things are only for our schooldays.

However much you may not be a poet, you might find yourself near to being still, a kind of Wordsworth for a time. You might get a sense of consummation, as I did, or nearly did, by the sea-wall. And this, I suspect, has something to do with an *order* 'far more deeply interfused' Natural laws interfusing with the things of sanctification?

I never told anybody. Such things seemed to belong to that time and place. But I found a scrap of paper in the Himalayan Temple that had almost been Ulli's final home; some lines scribbled from memory, with the headings: 'Wordsworth - Tintern Abbey'.

So, perhaps he had remembered that conversation of ours in the kitchen . . . Wordsworth . . . listening backwards to his childhood, possibly for that interfusing. Maybe Ulli and I had both been listening, joined by something - a sense from the time before we had put our childhood away.

I hadn't known then - when something like intuition had surprised us both, and I had quoted those lines, that in the sweep of time, somewhere up in those mountains, he'd find his Xanadu: the consummation of obedience, of 'following his star', call it what you will. Following *our* star, even. Yes, I suppose I had a star, too . . .

Such things belong to that time and place; to his 'Sun rising over Tibet, while India lay under that full moon.' We've still got that letter of his.

I also sometimes can believe thoughts that seem to come from nowhere . . .

How did you guess that, Dad?

7

And yet Ulli had been an extrovert child. Stuck up trees and couldn't get him down, sort of thing. Once he was stuck in a half-built sewer. Every child will play with cement if it's around, and it always seemed to be, so he would emerge from the caravan to dive into it.

Probably the beginning of our mutual understanding was that the two children had a little heap of cement to themselves every time I mixed any, and that Ulli could crawl and fall about amidst the lovely joists and other excitements of a half-built house.

At nine years old we sat him for an entrance exam to a local minor public school that had, as a Free Church foundation, a very strong independent and free-thinking tradition. Minutes before the exam I lost him - playing football - but he still got a free scholarship.

On speech-day he was in the PE team; week evenings on the trampoline; hated running, and half-hearted about rugger. A long, loping gait, long hair - the washed hippy type. About a quarter as much pocket-money as many of the boys. Too many kids get spoilt these days.

I am afraid that a lot of his interests were a bit half-hearted, more so as he got older; things that one had nursed in one's mind tending to fall by the wayside - an elaborate, magnificient, but abandonned chemistry set - a present from an optimist who also couldn't be bothered - still litters the top shelf in the garage; there's the kit for making a boat, specially bought in London, lying around in a box next to an old, but workable, steam engine, that I somehow can't throw away.

Then there was the months-long niggle about when was he going to fix his rear light, which turned out to have a blown bulb; too excrutiatingly difficult to have a look at. And so on, through all the now-loveable catalogue of tasks and duties that never got done because of lack of something-or-other.

Not exactly purposeful - but, dare I say, why should he have been?

Somewhere about the age of fifteen, though, there had been a patch of plodding, persistent hard work to get his O levels; a real interest in Scout camps; and he'd taken one or two holiday jobs over in Germany. But these ideas had started to flag as civilization sowed its seeds of unanswered questions. Ink splodges on the (still unchanged) red dining room wallpaper, and luridly different colours in sloping stripes across what was once his bedroom, seem to be almost archaeological remains of the past. And I think they mean something to all of us because nobody's said anything about moving them.

He became a monitor but was defrocked, or whatever it is, because he 'hated dropping other boys in it'.

From early on, his real ambition was to be accepted into a group of lads who were a little older than he was. They congregated nightly within sight of the window behind which he was supposed to be doing his homework. They were a close-knit little group, somehow a bit exclusive. I had the feeling that, whereas there were indeed hangers-on, one had to be specially accepted to get into the inner circle. Motor bikes featured noisily in their activities, attracting some little attention from the local police. This group became his life. I was impressed by their loyalty to one another; and a love-hate relationship between me and them fed upon that, and the noise of their bikes.

Although, in fact, he was quite good with nuts and bolts and with other people's bent machines, it seemed that human nature absorbed him more.

It's best I think to draw a veil over the question of motor-bikes versus homework. I can remember the anxieties, but not much about the homework. He got chucked out of the chemistry lab, but got an A level in that and three other subjects. I suppose it only means that the potential that the world requires was there. It wasn't sour grapes that turned him down the road and up on to the Ridge those nights. Only more recently had we at home seen how this so young meditation was to point to a star that the world couldn't see, one that was to dominate his life.

There was a streak of the Samaritan; a streak of rather dodgy Robin Hood (about which I got to know, but 'didn't know') like driving stranded friends, whilst under age, through the Dartford Tunnel. Things that came out in the wash.

There was the party that got out of control - a crumpled heap on the sofa - a sort of prodigal. Except that in the Bible he didn't get a drop of whisky.

He phoned me one day when we were on holiday and asked if he could use the house as a temporary refuge for a young couple. It turned out to be an unbelievable number of pints a day young alcoholic of good family and strong personality, in love with a near-psychopath. She (and I suppose Ulli) more or less got him off the drink. The lad had rescued her, I think, from a hair-raising past; unfortunately, her nearest approach to domestic science was throwing bits of waste and rubbish out of the window. I tried to be kind and tolerant, but also threatened to bundle her off to the doss-house. Sometimes I thought that that language, and Hell's Angels, was all she understood. Her more or less permanently demanding association with the latter having been colourful, to say the least. Other times I felt she was now experimenting, almost for the first time in her life, with reciprocal kindness - as if that was a new idea to her. It gave her a flickering, pagan warmth that intrigued me. But eventually I conned a nasty Rachman-type landlord into taking them over. I visited them once, and found her more or less wiping up the cat's mess with what proved to be the tea-towel, and I went home and had a bath before I could eat anything!

But I liked the chap; and Ulli certainly had his heart in the right place, even if he did rather jump in at the deep end.

Better than a £238 gravestone is the memory that when some minor matter kept me in hospital, shortly before Ulli finally left for India, he turned up to visit on eleven evenings. This usually meant a long detour into and out of central London, because he had a job towing workmens' site caravans to and from building sites. There were only two evenings he didn't come. When he did, there was obviously a parking problem: a jeep and a big caravan right near London Bridge. The admittedly rather anti-social solution worked beautifully. He dumped the caravan somewhere on a widish bit of pavement - as if they were going to dig up the road the next day. I suppose that everybody, including the police, accepted this ghost gang without question.

Sometimes there are advantages in parents being rather on the easy-going side.

I tried to tempt him from the pot that drew him down to the beach at Brighton, after dark, with other youngsters, and to interest him in the idea of living on a friend's sailing-boat. Sailing out of Dartmouth we lost the rudder near those very rugged cliffs, and the lifeboat asked the Navy to stand by. On another occasion, when we were moored up because of a gale, the rubber dinghy was blown up in the air, upside down like a kite, and a nude lady swam round collecting the oars (and collecting cheers). But these attractions gave way slowly to the half-hippy call to go East again . . .

He had already been stuck in Iran for months through 'irregularities' (nothing sinister), but got smuggled out. He had then tried to settle at Imperial College, London with lots of A levels; but six weeks in Notting Hill only seemed to heighten this restlessness. You sense there's something wrong; but what do you do when your son, nineteen years old, who's got involved in no end of trouble, comes out almost in one breath - but one sincere breath - with:

"Dad I want to leave and go. I want to become the holiest, purest, goodest man in the world".

Sitting in a pub, what do you do?

I knew that he and his friends hankered after Zen Buddhism; read books with strange titles, and so on. But this startled me. I tried to make allowances for youth, for naïvety, and idealism, perhaps for over-confidence - but the conventional, embarrassing, reactions seemed pathetic, hypocritical. I remembered the fruitless, suburban desert of my own youth, and my own later wanderings. In particular I remembered Henry Drummond. What do you do? Perhaps I went over to fetch a pint, to try to digest all this. To wonder that he had treated me to (or trusted me with?) such an unusual outburst. I needed time. It was no more typical than that poetry incident.

But if you'd once let something like Henry Drummond soak into you, you couldn't really moan. If you felt that you had found your own answer to life, you'd want him to find his.

I said he'd better go.

His career? Well, God knows what would happen to that! But then, if you hadn't sat on that gate and opened that faded little book at that page, you'd probably have agreed with all your friends that he'd been too much of a nuisance already; and priorities could well have been fixed the other way round; and God knows you'd have wished he'd keep these things to himself . . .

And I suppose, that when you got to him - sick in the little Temple (most of us would have got there somehow) - you would never have found that you'd taken with you a big part of what he'd really gone to look for.

You'd hardly credit that that run-down, beery old pub meant more to us, Ulli and me, than those tall Imperial College buildings. Not that we ever spoke of it. But, in a way I'd like to go back there. Something more than smoke and the smell of beer could be holding the ceiling up. It was a sort of casting-off from a spiritual suburbia, towards those strange echoes between the end of some Indian rainbow of his own, and the end of mine, so long ago out there. Perhaps that echo had cradled his future through all his early years? I was to feel it standing by somewhere at his death.

And so he resolved to go East again.

Eventually, fed up with working on a tiny freighter to get his fare, he set off back through Iran (a chancy idea), and ended up making his way through Afghanistan and Pakistan, via bribes, and under blankets, into India.

I suppose you'd call it trekking; weeks among the snowy and icy remoteness of the far North West of India - 16,000 feet in plimsolls; grunting bears; and a razor-like edge, where the sun rose over Tibet while India still lay under that full moon. Which all ended with him destitute, picking amongst the rotting fruit of village streams, but eventually finding his way to a little Hindu Temple, where he stayed a year - becoming a sadhu, a Hindu ascetic, devoted to the guru; amongst, as he put it, 'far-out Mahatmas'.

More and more concerned with starting up a kind of Temple-based East/West Oxfam, to bring some understanding between the continents, he wanted me to go out and join him in 'stirring up the lethargy that seems to choke any initiative out here'.

But suddenly a telegram arrived saying he was sick. Which landed me, woozy from injections, in a place called Chandigar, north west of Delhi - or more exactly, in a hand-drawn cart, two gharri-rides, an unintelligible argument and two bus-rides beyond it.

The hand-drawn cart was a two-wheel affair. Once I had been bundled into it, my "Pandoh-Manali" (the name of the village near the Temple), produced only a blank stare from the native dragging me along. He conked out. I got out and pushed.

The bus he eventually pointed out stopped in villages where rotten fruit clogged the filthy road-side trickle. I remembered the word for a plant that feeds on decaying organic matter: Saprophyte. The word seemed to hang over and describe the whole village, not just any plants that could have survived the place. I tried to shut out my imagination, and swallowed some tea and rice. I do wish they'd boil that milk! I remembered how Bombay milk used to arrive in urns with a handful of filthy stable-straw jammed under the lid to keep it on. We had received a letter saying that he had been living on 'fruit they couldn't sell'. I wondered how ill he was.

My mind latched on to a vivid picture of some diseases that end up with a blocked air passage. Perhaps he would be just like that, going blue. The words of a doctor friend came into my mind: "You put a pen-knife under the Adam's apple, quick. It's better than letting him die" - it haunted me, the idea of this surrogate doctoring; could I do it? - "Don't worry about the blood. Save his life".

Those antibiotics, smuggled from England, would stop any infection.

About 9 o'clock in the evening, after one hundred and fifty miles or so, the bus (now empty), arrived at Pandoh. Apparently this was the terminus. I got out. By the lights of the bus I could see that I was standing on what appeared to be a low mound of earthy hardcore. I could make out a few wooden huts nearby. Strange! Where *were* all the villagers? Not even an idler! No touts. How far had I to lug my bags? Where was the Temple?

I asked the bus driver if there was a taxi. "No taxi, Sahib". He and the conductor seemed quite unconcerned, and I lost my temper. I held up my wallet and banged a fist on the side of the empty bus. "Bus Taxi! Bus Taxi!" I shouted. They suddenly grinned. I jumped back in the bus and it started off, unscheduled, down the road - and I had longed for London Transport! After about a mile it pulled up. Suddenly there was a crowd of villagers, this time all helpfulness. Two of them grabbed my bags, and we started clambering on all fours up a rocky path towards a light on the hillside.

The Two Himalayas

I found myself in the little garden of a small wooden building that had the distinction of a gabled roof. The door was open and some half-dozen Indians came out into the porch, somehow engagingly agitated and chattering. I saw that they had been sitting round a low wood fire. I gathered that this was the Temple, and it seemed clear, from the excited, welcoming faces, that I had been expected. But where was Ulli, my son, the patient? Where was the sickbed? Whatever was I to find?

As if to answer my questioning face, one of them came forward, a skinny priest/pupil type with an ecstatic expression, and for a bewilderingly long minute he put his arm around my shoulder. I didn't know what to make of it. I didn't quite like it. Then suddenly, I thought that this ecstatic expression, this would-be blessing of the young priest's comforting arm, was maybe their way of conveying to me that my son was in Heaven, their nirvana maybe.

I wrenched myself away, and asked bluntly: "Where is he? Is he dead?"

The young, shaven, monkish figure said "Dad" - and I recognized my son's voice, then his hands. The rest of his body was unrecognizable - so thin and wasted.

I went into the Temple shaken. There he was, excitedly prancing about and saying that he was well. Not the desperate sick-bed scene I had imagined all this time, but a desperately thin body, tensely overactive, and enjoying the talk of illness and the telegram as a huge joke - happy that the telegram (sent unbeknown to him) had succeeded, above all, in getting me there.

To be 'joked' half-way across the world on a spurious sick-alarm crisis was initially infuriating, but I tried not to show those feelings. The emaciated body was not spurious. His weird disregard of his ghastly appearance took my mind off the first surging thoughts of "what the Hell!" was going on; and the hospitality of the East was too powerful, too reconciling . . .

The guru, Baba Dass, gave a good impression: very handsome, radiating a welcome through the language difficulties. I gathered he had been worried about Ulli's thin condition. He kept repeating, "Mind confused". So well he might; but what horrible meaning lay there? Ulli, half Temple servant, half my son, just laughed this off. What could I do but keep my cool, be pleasant - not spoil the obviously friendly atmosphere - and try to figure it all out?

One or two villagers had a smattering of English, but the guru's, seemingly, was confined to only two words, "mind confused", which gradually induced doubts; and the fear grew and crystallized around them. Ulli's near ecstasy at seeing me had indeed too much of a touch of insanity. Was it only excitement and religious fervour, or was it (to Western eyes) nearer to madness?

Amidst the confusion, and far into the night, the thought of a son who had become unhinged gripped my mind. A thin body, strange Eastern ways and clothes - these things I could bear; but the fear of a mind that had broken up kept me tossing most of the night.

Meanwhile a wind from the Himalayas blew up and rattled at the corrugated iron sheeting of the roof. It seemed to me a devilish wild orchestration, hammering in the same tormenting message, until, I suppose, I fell asleep.

But slowly I sensed that you couldn't quite judge it by Western standards. Down to six stone, I guessed, presumably through crouching with blackened hands over the fire, and lost in some mental pilgrimage night after night. Apparently on a "four potatoes a day" fast, trying to be a sadhu too quickly? And then perhaps, home-grown pot on an empty stomach... What else could one expect?

I grasped at the idea of hallucinations; and a quiet admiration for these wandering sadhus put such a hope in its own special context. These strangely relaxed tramps, apparently drifting from Temple to Temple, suggested profundities rather than inanities. Dirty, by our standards, their mental superiority to cold or heat, food or hunger, and to their rough sacking, seemed to carry a background reassurance that was somehow infectious.

I began to hope; and decided to feed him and wait, and to think more about his 'Tourist Bungalow' - knowing that it was more for an understanding Oxfam than for tourism.

Worries slowly receded into the Temple atmosphere, and gradually sorted themselves out into something like a very thin normality, but within the background of my uncertainty and concern about the 'influence', if any, of pot.

Ulli told me how a rather disturbing young girl had arrived some while previously, and left again; making it all a bit more challenging, a bit more complicated. He said she had stayed for about a week, during which time he had admired her little attempts at quite complicated English, and those more difficult attempts to try and explain her journey, spurred by an unusual female wanderlust that somehow answered his own nature. A friendly interpreter had spent two or three evenings in the porch over the fire, and told about other things in her story. We talked a lot about that.

He had begun to suspect that, secretly, her vulnerable young heart was getting lost in the wonder and despair of a fairy-tale situation, which wouldn't quite leave him either... Coming at night-time... A situation that was no doubt to burn cruelly into all that made her different from those hundreds of girls who never thought beyond the arranged marriage future, or lack of future.

He wondered whatever would happen to her in the next few villages. Whether this 'uncle' she was going to visit near far-away Rishikesh would be kind, or just reject her; the 'Rishi' probably had nothing to do with her name, Rishka. He wondered, even, whether that uncle's invitation was a made-up story to make her journey more acceptable to wondering villagers on whom she might depend for odd jobs to buy some food and keep herself alive.

And he told me how she had suddenly left. How a little tear had escaped, which probably alone had told the truth.

Some few days later he had got up and told the guru that he was going north, up the Kulu valley, to try and find out if she was all right... if indeed he would ever be able to find her. Kulu was quite near, and he had a hunch that she might well have gone there first, rather than banishing herself to the cold-seeming distance of Rishikesh.

But what he found was a cold Temple stone floor on which he had to sleep. He felt fairly convinced that she had called there offering to gobar their floor, which would have made it warm; but for some reason they had rejected her - presumably because girl sadhus (if such there be) are highly unusual. This meant that it could only have been her he overheard them joking about, which didn't seem to offer too much of an outlook for her travelling.

The Two Himalayas

He had shivered through the first night, then - a breakdown on the buses - shivered through another; so that by the time he had crawled back to Pandoh he felt thoroughly ill. This, seemingly, was how his illness and loss of weight started, the 'best cure' for which was the stupid 'starve yourself except for potatoes' regime, which clearly had not helped. This is, of course, where I came in, found him so thin, and tried to get him better.

During that time another village girl came one morning to do the gobaring of the Temple, and I watched her with some fascination, spreading cow-dung with her hands; an odd mixture of a special kind of nubile humility. Which made me wonder what I would have done in Ulli's position. But she was off somewhere, leaving the gobar to dry, and me wondering if I didn't secretly rather fancy her. Perhaps she had been the only Indian girl whose knees I'd ever seen?

Sadly, I couldn't seem to remember whether she kept them dry, poor thing.

Some time about then, amongst other things, I remember puckering up my mind at the mention of this Rishka's father's name, but I thought no more about it.

8

Rishka had set off for Delhi and beyond, leaving the village, the 'criminal tribe' settlement that had been her birthplace and her home.

For her fellow-villagers it was as if one of their people had set off for the moon. Not one of them could fasten on to the idea that a girl - a girl of hardly twenty - should board one of those big Bombay-Delhi trains by herself, so flying in the face of every convention or custom, of every idea, of a woman's place embedded in Indian village thought.

It all worried her, frightened her, and somehow saddened her, because her two parents had been half with her in her decision: half sad, half understanding. They were a little proud, too; both secretly knowing how much they had felt, and they sometimes talked as if the idea had been their own. Maybe this was to make up for their earlier desire for a boy-child instead of a girl. Yet this reaction was one that neither of them could quite justify to themselves, and was in any case quite unheard of.

Lakshiman, her father, friendly with the local station porter, wangled a free ride to the next little station and back. That had helped - taking some of those awful last minutes from her shoulders.

I can imagine quite vividly, how, as she sat back in the hot, bumpy little carriage, now at last alone, her mind swam; bumping, as it were, from one rather terrifying thought to another. Here she was, alone, a girl who'd never been to a city before. God! what she was doing?

She said she'd fallen off to sleep with exhaustion, and was only woken by the train slowing down and bumping into a station. She'd heard the name before, which comforted her; but wasn't it supposed to be a Sahib's place, where they came when escaping the heat they didn't like? And that sapped her confidence, what little she had, even further. Only very, very occasionally did any Sahib appear at her village. Some were supposed to be nice enough, but others gave her the creeps and would leer and make frightening talk.

Delhi would be full of Sahibs; full of rich, rich Indians, caring no way for her as a person, nor whether she lived or died. So she had been told, anyway. Trust any of them? She didn't know.

And where was she to sleep when she arrived that night? Again she didn't know. Sleeping in the open air was second nature to her - but in a big city, with people, perhaps nasty people, crowding the pavements? Beggars, nasty, creepy little alleyways, dying people even? Oh God... what would happen? What could happen even this very night?

And yet, I had the impression that in spite of it all, she felt, somewhere deep down, that she *was* different. Her mother was different, wasn't she? - speaking a little English,

The Two Himalayas

from an early childhood that was lost in some family mystery. But somehow a little proud. Thinking many of the same thoughts as the other villagers, and yet different. God knows what the story was, but her mother had inherited and nursed this 'English-speaking' idea so that her daughter might grow up and not be condemned to this village poverty - to menial jobs, like spreading their primitive village floors with cow-dung - but to something a little better. And when she, Rishka, had helped at the local hotel she had spoken her first few halting words of English because the lady there was helpful and amused all at once, and had encouraged her.

And her father? He was somehow different too. Once or twice he had spoken of the English Sahib who had, for some unknown reason, taken him under his wing when he was an orphan boy of ten or eleven. Who'd given him a piggy-back, instead of three annas for lugging a heavy case up the hill. A long time ago now; but her father remembered taking the Sahib to see the only person in the village who knew a bit of English - a young mother with two or three children. He remembered how they had both gone back to the squalid little hut several times, because the Sahib wanted to try and get a cure for the two tiny toddlers, who were crawling about with some nasty-looking, unknown, and rather dreaded skin affliction. He was to remember the details of all that because, as he grew up, he found he gradually became more welcome there than in any other hut; and because, eventually, he had married the elder daughter.

So her father, this Lakshiman, had grown up with an idea about one day going to England. The Sahib had even talked of that. He'd had some idea of taking Lakshiman there, together with a big, tall Indian man who knew lots about books and such. But it had proved impossible - too hopelessly complicated. The Sahib had wondered a lot, he said, about staying on in India after the war, but had thought up this idea instead.

And as he, Lakshiman, had no male children and never managed any travelling himself he, in a way, wished this upon his daughter, who was now sitting very worried upon this train. Very worried, except that her father had also grown up with a very precious little envelope, given to him long ago by his Sahib, which was some kind of 'reference' to help in getting a job, which he'd never been able to use, but which he had secretly been keeping for his daughter. This might very well get her a job, or give her entry to Temples, or whatever. And in this little envelope had been a nice little present of a few precious rupees that Lakshiman had saved for his daughter. Not that the fare to Delhi, or further, would cost much. Not third class anyway.

But third class or no, I suspect she felt 'different' enough just by being on a train, going on a long journey? The thought gave her confidence and made her feel secretly pleased that she wasn't ugly. She supposed that only richer girls thought about that, girls who lived in houses and looked in mirrors like the Sikh doctor's in the village, whose inside room had one of those shiny glass things that she'd wanted to get a bit nearer to when she was gobaring their outside 'backside room'. I smiled at the thought.

I suppose she was soon able to puzzle out a few things - about knives and forks for instance, and other things inside houses? Her idea, of course, was to get a job somewhere so that she could solve the sleeping problem. No doubt she'd work all day not to have to sleep on those pavements or in those alleys, provided she could get just a bit of rice to keep alive on.

I remember she said that, quite a long time ago, she had been to Bombay for a day with her father. That was more a frightening than a comforting memory; her world had

The Two Himalayas

only really been the local village with its few houses, one or two shops, the so-called 'hotel', and its one-eyed police station and village railway station to give it some kind of official existence.

Bombay had quite paralysed her, and so probably would this big Delhi place. Wherever, for instance, would she find a friendly woman, like the village doctor's wife, who would laugh her in her kindly way into trying out her English?

In all her confusion, she thought mostly towards that possibility and tried to get her hopes there ... until the train pulled in to one of the first suburban Delhi stations, and then started to crawl towards the next. Thousands of pushing people at each station. The railway embankment a kind of public lavatory half the time - but mostly where it was sloping down the other way.

Having been brought up in a village of the 'criminal tribes', she wouldn't of course have quite the European outlook on these things. But I somehow suspected she would have had a kind of natural distaste that carried an uncomfortable foreboding about Delhi that she couldn't understand.

She disliked the pushing and confusion of another kind that took over when the train finally stopped. It had only one advantage; she couldn't sit down and just give up. She just got pushed along, with a kind of despairing instinct not to let go of her few bits of luggage. So that in the end she found an empty seat on the station, and held all her rather pathetic-looking belongings beside her. It didn't seem unexpected to her that just next to the seat there was a very small child, sitting naked on the ground, eating into a little heap of curry that I suspect would almost drain a European's unpractised tear-ducts or burn out their throat.

For a time she sat there, assailed by strange worries and disturbed by all the passing unfamiliar sights, while somehow she tried to nurse her thoughts towards some kind of confidence. This certainly was *different*, and she supposed that, in a way, this was what she had wanted; this restless urge (inherited was it?) to be different.

But in this rushing, over-busy world of practical living, of just keeping alive, I imagine that she shied away from a memory of those secret night-time thoughts that crept into her mind now and again and, that she knew were somewhere mixed up with her being here anyway. I tried to put myself in her position. How on earth could she ever remotely hope for such things? Dreams of one day meeting a kind of young Sahib; yes, even a white Sahib, who would be a bit like the one who brought that strange longago kindness of her father's memories.

That was a dream-world. And all around her, surely, was the opposite of that dream. Except that a little wave of kindness drew her toward the little child, and she rummaged in her bag for some morsel of chapatti, or the like, and gave it to the child, who looked up and gobbled it up in surprise. The young mother - children like that do not always have mothers - saw this, and smiled across at Rishka, and they talked a little; which made Rishka feel very much better. Rishka learnt that the best thing she could probably do for the night would be to walk towards the big square with big street lamps on all night, where there were always lots of people moving around, and where she'd at least feel (and probably *be*) safer, even if not very comfortable.

And that's what she did, and felt grateful for that stirring of those not-so-heartless ripples at the edge of all the rushing.

The Two Himalayas

The next day or two didn't seem to achieve very much. She was in truth a bit scared to go on further into the unknown beyond this Delhi; and also not a little scared to stay there. She slept without too many anxieties as to those more dreadful imaginings, in the same big, brightly-lit Square, and wandered rather vaguely about during the daytime, secretly trying to pluck up the nerve to ask for some kind of little job - no matter how menial.

She found that parts of the City were just one big jumble of every kind of humanity, every kind of beast-of-burden, pulling every kind of ramshackle wheeled contraption. The streets, especially the narrower streets, were in a state of permanent shouting confusion that defied any kind of order.

Walking at the side of one of these crowded roads, on about the third afternoon, she suddenly felt a blow from some hard object on her left shoulder and fell half forward, half sideways into the roadway near the gutter.

Ulli had spent hours talking to her about all this, and when he told me it was as clear as if he'd been there. She heard violent and angry shouting, and had looked up to see a young white man standing up in the gharri and wrenching the big whip from the driver, while another very agile fellow had leaped out of the vehicle and was grabbing the horse's muzzle. It seemed that the driver had lashed out at his horse, poor thing, to make it get as far from the incident - the girl on the roadway - as he could.

But it seemed also that the two young men weren't having that. They were in some kind of uniform with the blue peaked caps that, she had been told a year or two before, were from ships that came from other lands. One of the men came over to her, a little bit embarrassed, but wanting to find out if she was hurt or needed help; taking to hospital even. He was a bit fat, but with a kind face, and spoke in that now-recognizable English, but much too quickly for her to understand. So that the little bit of farewell advice from the doctor's wife came to her rescue, and very shyly she got out the words, "Speak slowly please". Whereat the young Officer smiled and did just that. And somehow - the other man having roundly cursed the driver, telling him I suppose to get lost and not to start on about any fare, or whatever - somehow they reckoned the best thing they could do would be to take the girl back to the little hotel where they were spending a few days, and see if they could get her somewhere - anywhere - to sleep for a night or two. The idea that really quite an attractive, and so shy a girl had nowhere to sleep would have appalled them, and a kind of chivalry, not too prevalent in that hot country, brought them into the hotel office asking about doctors, about some kind of kitchen job, some kind of way to earn a night or two's shelter. Maybe - may very well be - a rupee or two got smuggled across the counter; but the owner turned out to be a kind man, who seemed to like the girl, and said he would do what he could.

Poor Rishka had felt terribly embarrassed, but also very grateful. Neither of the two young men was, in any case, anything like the young white man of those secret imaginings. They were too hefty, and perhaps a bit on the noisy, restless side for a shy thing like her.

But it was - especially when her bruise had healed a little - the sort of lucky break that she had never dared hope for. The hotel keeper's wife, rather unexpectedly, was not unlike the doctors's wife at home and seemed to take kindly to this quiet, village girl with her hesitating English and willingness to do her very best at all those little cleaning jobs that plagued a hotel keeper's wife.

And, it was quite unbelievable! when one of the attic rooms had to be repainted and cleaned up a bit, the lady said she may as well sleep there for the night - giving her an old mattress, that felt like heaven after her village life in those huts and tents she had grown up in.

More unbelievably still, there was one of those big glass mirror-things, where you could stand and see yourself from head to toe; and even a bit of a crude shower - to attract the white travellers who seemed to be used to these things; the water was cold, cool rather, being up at the top of the building. But in that hot country, this was hardly an affliction.

Hardly an affliction! In fact for her it was all an irresistable temptation; and for the first time in her life she saw herself, her young figure, daring to be seen in a mirror for just a minute or two; hoping desperately that everybody was fast asleep and that there wouldn't be a tap on the door to "come please, and clean something". Most unlikely at well after one o'clock in the morning, but still a bit worrying. And that shower. She wondered. Surely the couple who owned the place slept right down on the bottom floor. If they *did* by any chance hear water running how would they know it came from her particular room, and not from one of the seven or eight other guests?

Back in her native village there had been the little stream that ran from the hill where her 'tribe' lived, to down near the station. She knew, you see, that the small boys of the village used to play about in that. The Sahib, a long time ago, had given them some soap, which they'd never seen before and which caused lots of exciting bubbly stuff. So they had said. But not for the girls. Not that sort of thing. Her nearest thing ever to a bath or a shower had been from a little urn inside a hut, with soap only once that she could remember. Nicer to be a boy, she thought.

Gradually she must have plucked up courage. Was it really very bad and naughty? And then she was underneath it, and had turned the tap thing and got it running. The first time ever. How lovely to be a Sahib, a Sahib's wife - and have this every time you felt hot! And of course her long and really rather beautiful shiny hair got all wet, and she tried to soap it, and wondered how it would look when dry. That was a point. It would take all night to dry . . . and the man's wife would surely see it. But she was somehow mesmerized by this, her own hair, flowing over her shoulders, over her never-ever-seen-like-this-before body. Dripping wet, she smiled into the mirror to keep this picture in her mind, perhaps to give herself just a little tiny bit more confidence - confidence that she needed so much.

She began, though, to lose confidence about what would happen in the morning, until she suddenly realized that the old cold-weather sari that the doctor's wife had given her as a little parting present would surely do as a towel of sorts. So, when she went rather hesitatingly downstairs in the morning, nobody said anything. Perhaps the manager's wife *did* glance once or twice towards her, but if she had noticed the new and shiny appearance, she was kind enough just to keep it to herself. Just a little smile, was there? Could that have been a hint of approval?

So she stayed there really quite a long time, learning lots of little things about houses, and the posher world of washing, and eating nicely; and lots of cleaning; and quite a lot of English - at which she got better and better.

But eventually she started wondering whether it wasn't time to set off again for those northern parts that had really been the object of her journey, and to leave this big Delhi to those who only wanted towns. Confidence-wise, she was quite a different person and it no longer seemed such a frightening prospect; after all, her father had given her the fare, and she had this 'reference' thing, didn't she?

The urge won, and she decided to go - but first to Chandigar; perhaps she could stay there for a while. Knowing about hotels, couldn't she perhaps get some sort of job again?

Which in fact she did. And here, secretly again, she managed to wash her hair and feel all beautiful once more.

I suppose that, without realizing it, she had become almost the only Indian female around who risked the kind of journeys that foreign boys or men seemed to find so necessary to fulfil their inner urge. Young men who left England and ended up in places like Rishikesh, where there was a kind of holy place - an 'ashram' they called it - which her father had once dreamt of seeing. Perhaps she ought to try and get there (or somewhere like that), so that if they did not reject her as being female she could one day tell her father. He would have cried with pleasure. But she did not quite know why, nor why she thought that.

There was of course another journey to be paid for to get out of Chandigar, this time by the shaky local bus. They never seemed to have any spare seats - or so she heard from a girl she had chatted to in the ramshackle old hotel in Chandigar, where they had worked her ruthlessly for a few scraps of food and a place to sleep in the backyard.

The idea of calling at one of the Temples that had gurus and took in wandering sadhus intrigued her. Males only, it was said. But she didn't quite see why she shouldn't offer to gobar their floor, or the floor of some not-too-holy room at the back somewhere, where she could spread the not-too-holy cow-dung for not too many annas. [The cow is, of course, a sacred animal].

The first one didn't take to the idea.

That was in quite a sizeable town called Mandi. But there was a religious festival on that day, with a lot of cheap home-brewed drink about, so that people were a bit more talkative than usual, and someone suggested she go on the bus to Pandoh, not far away, where there was another Temple, and a bit more money about because they were building a big irrigation scheme.

After a worrying, horrible night, wandering about in the aftermath of the Mandi festival and trying to keep well out of the way, she arrived at the Pandoh Temple on the hillside, not too far from where the bus finally stopped.

It was evening and she was frightened, wondering whether she wasn't being very, very unwise - risking being rejected back into the high street of a strange village that somehow looked a bit uninviting - especially with the darkness coming on.

Nearly crying inside - trying to stop it being noticed anyway - she ventured near to the Temple porch. She was wondering how close a young girl was allowed to that very holy-looking shrine that stood at the side of the Temple proper. Two Temple helpers, sadhus she understood, appeared after someone spotted her standing nearby, looking very hesitant and shy. One of them was a gnarled, probably fairly young Indian, with a kind expression that seemed to be meant to reassure her, who asked where she came from, and so on; the second was dressed in Indian clothes, but looked foreign - which she soon knew from one or two muttered words of what sounded like English.

She smiled inwardly, and so in truth did he; something of a special kindness came from his eyes, a special kindness no doubt mixed with the effect of not having seen such an attractive young girl for a very, very long time. A girl who must just have washed her lovely hair in a stream, he supposed.

No way, she felt, would she be thrown out helplessly and unkindly; but there it had to stop. Here was that world again in which she knew her place. A pretty girl - but a village girl - whose feelings were non-existent almost; one day to be finally disregarded, when some arranged marriage descended upon her; which would no doubt be her fate. All somehow epitomized in this little scene in the Temple porch, where her private dreams sank into the ground; her dreams of whether she could grow up 'different' coming to mean nothing to anybody, and this adventure-trip serving in the end to rub it all in . . .

So that, without really any hope, save perhaps for some minimal kindness, and thinking that this would best become her as a wandering - strangely wandering - young girl, she asked whether perhaps she could do some gobaring.

The lowest of social humilities, but which didn't bring out any sense of rebellion in her picture of her humble self.

She thought, though, that she saw a sad expression on the foreigner's face.

"You gobaring?" he muttered.

"Yes, Sahib, me gobar."

"Me gobar," he repeated to himself; and, "Good God! . . . but I see guru. Perhaps that's best . . ."

9

For some months before the telegram had arrived, we at home began to feel that Ulli was becoming rather heavily preoccupied with the Guru Baba Dass and that pet 'Oxfam' idea; preoccupied with absorbing this spiritual hero into the concept 'Oxfam', to create a 'Dassfam' project.

His dream of a new Temple connected with the 'Dassfam' tourist bungalow seemed to be knitting together the best of East and West for a kind of mutual help, and for the cause of understanding - and, for God's sake! couldn't I come too?

But this guru-fascination had, for me, been making the letters home a little over-philosophical. How much was from the guru - paradoxically unable to speak any English - I do not know; but Ulli had started setting out what I supposed was the Eastern cult of humility, in its many different forms, in many appeals for our understanding. I thought afterwards how very appropriately this anticipated the questions she would bring, this gobar-girl, "spreading it so that it would dry and be the floor for others".

These others - men, and only men - came nightly to sit around, forgetting their boredom and perhaps their privations, in evenings that were quiet; congenial to each other and to a meditative approach to life. Lively, often humorous discussions, sometimes obviously referring to Hanuman, the god to whom this particular Temple was dedicated. Nothing ribald; no drunken nor even remotely drunken voices. Rather, mutual awareness than noisy loss of inhibitions. All dominated by the strange personality of the Guru Baba Dass.

It became clearer, after I'd been there a day or two, that this pot ritual would indeed have raised problems; but I suspected that they would normally have them well in hand; until perhaps, who knows? ... the young Englishman in a weakened state, not used to it as *they* all are, found the pipe a bit strong and a bit tempting; and so cause disruption, and thus become a focus for other local feelings?

I could find no other 'weakness' or personal indulgence in the guru's way of life. He seemed impervious to cold, sleeping through the early hours long after the fire had died down, in a thin half-length vest and loincloth only, legs bared in the cold draught. He was indeed remarkably fit, strong, and lively.

I felt instinctively attracted to him, personality to personality. We were naturally drawn to be kind to one another. By virtue of his presence only, the little Temple household ran as if every action and almost every object was sacred. It was partly this that convinced Ulli that the man was privy to things that elude ordinary mortals, and that he might well have had a rapport with a spiritual world - a world that was certainly convincingly real to him. I ungrudgingly concurred in all that.

But Ulli didn't seem to see him in any sort of critical perspective. The guru was too much a luminous, almost simple answer to Ulli's own searchings - Zen, yoga, Buddhism, meditation, and so on . . .

To me it seemed as if Ulli, having waded through these difficult paths, had found a quieter goal under the strange guidance of the presence of this man who, paradoxically, had no way of conversing philosophically. I suppose that is why he was a guru - almost a mahatma, they said.

This yogi-teacher-guru could somehow *speak* through the act of integrating this difficult abstract area with simple daily living. And this seemingly created the atmosphere for Ulli's fascination with trying to recall New Testament memories, those that had strangely survived from ages-before RI lessons, seemed to find a very special place. Something infectious deriving from their simplicity, I thought. A unique harmony surely, esoteric searchings, daily yogic living and the New Testament. A palpable sense of God: a strange setting.

No wonder he couldn't at first settle to, or bother about, being 'sane'.

There were other puzzling things about this guru. He seemed to be alarmingly double-jointed, which added another dimension to one's sense of encountering a life that had stretched the usual parameters. And, moving on to more contentious ground, he was supposed to have stood, night and day, for 14 years without a break - never sitting or lying, tied to trees and propped up one way and another to keep him up.

Of course I dismissed that out of hand as absolute rubbish, but in the end I began to wonder, or at least doubt my initial cynicism. Agreed, the motivation would be there - the spiritual prestige - the appeal of mind-over-matter holiness to the Eastern crowd.

He was supposed to have arrived at Pandoh on the bus, standing all the one hundred and fifty miles, in a bit of a state; then, unimaginably, continuing the feat in a local derelict Temple.

A proving of self-mastery, a wish for adulation, a pure desire to turn people to his God? Or just a ridiculous stunt?

The local headmaster and a Sikh doctor were more than half-convinced - both Western-orientated men whom I respected.

And he had shown me scars on his shins, left over from the injections he received against the pain, while at night helpers had clustered round to keep him up . . .

I recount it for what it is worth. But, to get it into perspective, I myself am unable to stand around for a number Twenty-Five bus!

I was on safer ground, for a Westerner, when I suspected him of being after my money.

There was this talk of building a 'tourist bungalow' close by the Temple, where I and my friends from England could stay. Only he couldn't afford it himself, not quite. And they were continually on about their new Temple. Ulli was especially sold on this - but trying more to convince me of its virtues, of how it deserved my sympathy. The guru seemed to add something that was rather less than ingenuous - interwoven with friendship - but enough to trigger off a little signal that I might be a friend, and at the same time the goose that might lay the golden egg!

Neither did I go much on his relationships with the local villagers.

The third morning after my arrival a crowd of about thirty or forty village menfolk had suddenly appeared, wanting to see me in the Temple forecourt. They apparently

The Two Himalayas

wanted to know if I was taking Ulli back to England to see 'best' doctors. He would not get better here, they said, getting rather hostile to the guru, who became surprisingly agitated.

Amongst all the pidgin English, I sensed that they blamed the guru, and that they were hesitating to mention pot.

I tried to explain that I could not order him back, and that he did not want to go back. In England "father's don't give *hukm*" ['orders' - one of my few remembered Hindustani words]. I could not take him back so thin. Not, I thought to myself, at once so weirdly Eastern and at the same time unwilling.

But I got worried lest this sort of row might aggravate the nasty streak in the guru, that was in fact to show itself over a stupid demand for a written health-guarantee - to ensure that Ulli had no illness-problem after I had left!

Not then knowing the Indians' care for their sadhus, I had wondered about some unthinkable end.

So I puzzled what to do with this crowd. Suddenly I hit upon the idea of a strange bargain: Ulli's health and security in exchange for my helping with the construction of their new Temple.

By sign language, and more odd bits of pidgin English and half-remembered Urdu, I told them if, while I remained there, he got "fatter, fatter; legs thicker, thicker", then I would buy some cement. I would buy a camera in Mandi the next day, and if he sent me a "click click" [photo] of himself in a few weeks, not "thin, thin legs" I would send out more money for their Temple. If later I got another good, *good* "click click" of him, and of the Temple half-built, I would probably come myself, pay a little more, etc.

Smiles broke out.

I crashed in and said - with no idea whether the guru wanted me to - "Guru Baba says Ulli stay if Ulli not too much pot".

They got the sense, and smiles spread. Then they all trooped off, leaving the guru to recover from his almost trembling agitation.

I got a cheap and nasty camera the next day.

This Pandoh is in the Kulu Valley, which is far more fertile and therefore more prosperous than most of India, but its natural, rather wild, rocky beauty was being spoilt by rows of huts housing the workers from a quite impressive underground river diversion scheme. No doubt this is desperately necessary, and presumably one day these huts will mostly disappear or disintegrate. This once sacred river is to be partially diverted into an eight-mile tunnel to join another river.

I hope it will be left, diminished in power, but undisturbed by the beavering trucks and dusty industry of man; for I now know it to be that Alph, the sacred River that somehow joined our schooldays. [Sadly, Coleridge's 'Alph' is now called the Beas].

I suppose that the hillside will heal itself, and that all this disturbance will transform the aridity of those poor subsistence villages in the plains. To those in the little Temple it *was* sacred in spite of it all, and so it somehow is to me, and always was - although long ago I never knew why; I was just a schoolboy.

Yet it retains some of its magic even now, especially higher up the rocky slopes - and most will be restored completely - that magic which to me touches that 'Razor's Edge' feeling, but mixed with the call of sadness.

So that it is strange - and to me emotive - to read that this Kulu Valley, and the remoter little side valleys, were chosen in Vedic times by sages and saints for meditation and prayer.

That it was once the southern limit of Kubla Khan's Mongol empire has something to do with its timelessness. Not very far away is Ladakh, sometimes called 'Little Tibet', the unspoilt remnant of the vastness that once looked down at me ... that had something to do with *my* past.

Even the little town of Manali, at the end of this valley, was pretty remote for my feelings. It is surrounded by 14,000 foot peaks, and is very cold, the snow not yet melted. Somehow, ominously, this is the start of what is said to be the highest mountain road in the world - from Manali to Leh, a military outpost and the capital of Ladakh.

At the time I had never heard of it, and it meant very little to me; except no way would I venture along the icy hairpin bends and horrible yawning drops of such a road to see it - or anything else!

I was spared the challenge. It is supposed to be - I did not know - that last and purest remnant of the traditional Tibetan way of life.

Conditions in more distant Tibet itself have we know changed radically since the Chinese invasion. A large number of Tibetans trekked through this grim-looking pass to their present wretched hovels around Manali. Presumably they must have trekked through Ladakh, which is scarcely able to sustain its own small population - its summer pasture giving way to wild blizzards during the winter.

Of course, it must be an intriguing place: monasteries clinging to spectacularly precarious remoteness, and taxi rides that aren't true - the strange vehicles wrenching your balance with the circus-act mentality of the drivers.

I suspect that nearly everyone destined to travel there must secretly want to look for a country where the only wheel was a prayer-wheel and the only flag a prayer-flag.

A country of nomads, of long caravan routes such as that stretching from Kashmir through Leh and on to Lhasa. A unique entity where there were no banks; where famous rivers have their source; where the dead are left to vultures as a symbol of charity; and where the Dalai Lama was utterly revered (and no doubt still is).

No doubt, too, there was intrigue in high places - though when someone was pushed out of a high-up window it possibly would be the ground which killed him, because of the edict against taking life ...

I imagine that modern Chinese political thought sits rather uneasily in these surroundings, where the most consistent thing has been the continued isolation from the outside world. Perhaps one could still find echoes of Ghengis Khan, and of his grandson, Kubla Khan, whom Coleridge enshrined in those special mists of our youth:

> In Xanadu did KUBLA KHAN
> A stately pleasure-dome decree:
> Where ALPH, the sacred river, ran
> Through caverns measureless to man
> Down to a sunless sea ...

Krupp managed to sell ever-improving shells and ever-improving armour to some forty-six nations and so, in a sense, had the world in thrall. Ghengis Khan did it, perhaps, more romantically, on horseback - controlling the Mongols and so overrunning empires from the Black Sea eastwards to the Pacific. He set up states, and their organizations lasted.

Kubla Khan, born to nomadic power and still a barbarian, eventually set up his capital in what is now Peking. Having conquered China, he adopted the ancient Chinese ways and learning of some two thousand years ago... Strange - there was a Chinese emperor all that time ago, who built roads and even canals - and who no doubt conquered someone else.

It was said of Kubla Khan that 'you could conquer on horseback, but you cannot govern that way'. In fact he made Buddhism the state religion, presumably not on horseback; for he abandoned the massacring habits of his grandfathers' generals, and established himself a Court, the splendour of which inspired Marco Polo, and, still later, the imagination of Coleridge.

Tibet was one of this Kubla's vast dominions; but Xanadu was his home - in south east Mongolia - where, I suppose, he dreamed far-off exotic dreams ... a pleasure-dome in that far land in his setting sun - where Alph *my* sacred River ran.

One cannot stay near this Manaii/Kulu valley without sensing something of the remoter drifts of history, but I grew more and more concerned with trying to puzzle out those 'Indian intangibles', and with how to fit in my ways of dealing with those of the guru, whom I found in some respects rather less than saintly.

I was worried because Ulli seemed to be finding himself in the rather painful middle, so that his earlier inspiration, well-founded in many ways, was sagging gradually into a kind of partial disenchantment: the disenchantment being where the Temple impinged upon the world outside.

My presence seemed more and more to have polarized the East/West divide. As a moneyless sadhu, Ulli hadn't come up against the questions that now inevitably arose. It was difficult to keep clear of the vaguaries of village politics, the rather shadowy police interest and those subtle 'intangibles'. Not least, and rather worryingly, I began to see that flexibility on my part would often involve too dubious shades of honesty for comfort.

And my concern about Ulli's health had been aggravated by my having to write that letter accepting responsibility for it - for any illness, I suppose - even after my stay was over, with me back in England.

I gave them a meaningless scribble, but it was not very inspiring; apparently it was to save the guru from having to bribe the police not to pin a charge of some sort on him, should Ulli succumb to some nasty disease. I couldn't quite work it out; and I didn't like it.

Then there was the time when the police suddenly appeared at the Temple, with seemingly little pretext. I had no idea what that was all about either, except that, a few days later, when we were calling at a teashop with one of the Temple sadhus - on our way to see the District Commissioner [a government official] - the 'tea-shop' turned out to be the front office of a police station.

First Ulli's passport was demanded, and then mine. I got hopping mad, and my inborn contempt for petty officialdom rose to the surface, so that I was almost grabbing my passport back when the door opened, and a decent, very gentlemanly, officer walked in and resolved the whole situation very pleasantly. I think it was their off-hand lack of any attempt at explanation that suddenly had made me feel that the only non-hippy type Englishman for hundreds of miles shouldn't be treated like that - very foolish, I suppose, but sadly, much of this is not very inspiring. Yet maybe, after all, it puts Ulli's eventual Franciscan decision into its true perspective.

The District Commissioner himself was, however, very pleasant. I had the distinct feeling that Englishmen who behave in a gentlemanly manner are very welcome and respected. The Commissioner was, after all, an inheritor of a traditionally respected line of men who had devoted their lives to the country, attempting to be fair in administering, rather than perpetuating an old boy network. It may well have flattered him that I should have asked his advice; but I felt that he knew, as I looked at the portraits of old DCs in his office, that I was expecting integrity. That I felt it was enough to leave any Ulli-guru imponderables to those old paintings hanging on the walls. But I also finally decided that the police subscribed to some strange guru angle, as intangible as the charisma of the portraits.

Only rarely is a European privy to the understanding of such things.

I mentioned that fascination with humility which had drawn more and more of the pages of Ulli's letters home. I now saw that he slept with bare feet and legs, his body wrapped in one blanket. Like the guru, he didn't seem to feel the cold - some kind of mental control. At home he couldn't get far enough into the fireplace! and I shook my head in puzzled marvelling, and snuggled myself down.

But one night he felt poorly, and from then on he had half the big mattress. Secretly I hoped that this start into comfort might wean him back nearer to thoughts of coming home; but secretly, also, I knew that he was happy where he was, as a sadhu.

Pondering all these things, I had started looking through the untidy roll of paper that he had tucked away on top of the old chest that took up most of his 'corner' of the little Temple room. This, apparently, was the 'Christmas Letter' that he kept promising us but had never posted.

It - the part I read then - helped to confirm the way my mind was running; a growing daily reassurance that I could discover a steadiness in his eyes and in his smile, and draw a temporary curtain over more and more of my fears.

But it drew me behind that other curtain - as sometimes in the London coffee-shops where I would occasionally read his letters - where he was out there stirring up the ghost of my old yen for India, its Lakshimans, and its crying villages.

Although I didn't read it all, I was glad that I had it there in the corner. It seemed like some kind of spiritual passport, a window on his Eastern pilgrim's progress. This castle was a Temple built from the East. Could it, unaware, have been facing the West, waiting for an unbuilt wall? Waiting for someone, looking for their 'Razor's Edge?'

It all seemed fairly reassuring, the Christmas thoughts; the sudden little compliment about how 'that most memorable moment of his life' was a confession one evening in our kitchen about LSD, and the comfort that this 'experience' was now being almost equalled in less dubious ways.

I get a little bit lost in themes such as the endowment of Hindu gods with numberless hands; but find myself again in descriptions of those 'far-out guys' and that (ecologically) unadulterated life, and - a little hypocritically - criticism of the slavery to possessions.

The Two Himalayas

And I sense a very real conviction that some kind of release from daily tensions came from his pondering upon thoughts of the immanence of God - the idea that God somehow spans all opposites; the desired result being that such things as heat and cold, and so on, that bug us, become less disturbing, less able to throw us. I can get that idea even if not too much of the immanence.

And so on, until some touch of youthful euphoria brings me hesitating again. Yet, his 'freaks here are top class, not bottom class as in England. I am a sadhu', means that Western judgements, and cynicism, do not fit.

I put the letter on one side, thinking of all the points it raised. How his hands 'were black with smoke' - which they really were, crouching most of the night bemused by all those strands of thought; how he was always kept on the go; how he'd taken to all those strange and not-a-little primitive customs, valuing that ecology of the Hindu gods more now than that of the West; and theories about money and possessions; his devotion to rebuilding their Temple, and his idea of our building that 'Tourist Bungalow'.

That bungalow kept on intriguing me. At a viable distance from the Temple (and, I must add, from that everlastingly blaring-out Hindu music) it could well add up to much that I longed for, and looked back upon, and believed in. Their unceremonial, hidden 'ceremonies'; their Eastern awareness. I wanted to be near - but not too near; and, if you like, clean, but not too clean.

It seemed (now that he had been accepted as a sadhu, an apprentice guru) that his object was to become something like one of the Indian 'godmen'. To live, so that every act is done in love to God and man; a kind of yoga, but not physical yoga, so absorbing him that Western medicine has to take second place to temple bricks. The surgery and injections being three hundred yards away, the bricks thirty miles.

But eventually I got him to the doctor, a kind man, who reckoned he'd got amoebic dysentery, anaemia and lack of protein ... "I think he'll take the tablets, but doubt whether earth for washing will be replaced by my bar of soap."

I shall always remember that surgery. I was taken from my modest place at the end of the queue, and given a special chair right next to the doctor - wondering not a little about the female patients in the queue. Most of them entered jabbering, hoping for the magic of a prescription. One man's trouser-leg got pulled up, an abscess was cut off with scissors taken from a bowl of yellow disinfectant, and out he went to wipe the nasty off his shin somewhere out in the field. Each time the bulging door was unbolted, and a female patient let in, I waited for embarrassment, but the situation never arose. I never quite understood. It could be that the Indian female would remain impassive whatever she felt, not having the status to get upset - possibly not even the status to be examined - I don't know. They haven't even the status to have or share an umbrella - getting drenched while the husband keeps dry.

Caught swimming once in the nude, I didn't know what they thought or if they thought - filing straight past, as if the jars of water on their heads paralysed every flicker of their minds. An odd experience, somehow filling me with sorrow. It didn't help me over the surgery; I had no idea whether I should have worried.

But it is such a different world, these sadly dirty villages. Some, like Manali where Guru Baba and I spent the Friday, are indescribable.

The Two Himalayas

But how was I to know at the time, shortly after I had returned home, what this extract from one of the recent letters really meant? . . . or might have meant?

'A day or two after a new little puppy had arrived, Baba the Guru decided for some reason to brick it up - I can only think to make it a more bad-tempered, savage, guard dog. He put a row of loose house-bricks under the front of the wooden seat in the Temple porch, so that the poor thing spent three or four days trapped inside this nine inch wide dark cavity, squealing pathetically. I got furious, and I used to put my hand down to open up a little gap between two bricks, look the Guru straight in the face, and let the poor thing chew my finger. And the atmosphere would thicken until I gave in - partly because the Englishman's code of animal fondness can conflict hopelessly with one's obligations as a guest.'

Later, as we heard from Ulli's letters, it had sickened and he had tried to nurse it. And Ulli had a sore foot, which could have let an infection in. Nobody really knows, partly because he was too casual at first, and too confused later.

I could probably have saved Ulli's life. But here there is a sort of double irony - interacting ironies that wouldn't be accepted in fiction: that we have to accept, and feel that they'd be too numbing to try and follow. 'Keeping as clear as I can,' I had written, 'of the worst horrors. Courtesy forbids too violent evasive action in this direction, so there is a risk obviously.' . . . Courtesy!

Ulli, Guru Baba and I went to Manali up the Kulu valley for two or three days - a none too clean experience, unantiseptic beyond belief. Sleeping in Temples, sad and dejected beggars sitting about, sharing their squalor. God knows what disease(s) I might have caught. I took tablets. Ulli and I walked five or six miles up a hill - equivalent to climbing Box Hill. He seemed fit in that respect and I believed he was putting on weight. He was probably now eating too well. I was OK, but had plenty of discomfort, long starvations, waits, frustrations, etc.

Guru Baba had turned this trip into a Hanuman music/preaching circus. He took a huge wooden box full of broadcasting gear, which two men had to lug from Temple to Temple. One such - way across some paddy fields - turned out to be a derelict ruin, with no signs of friendly, tea-making inhabitants. The other two, more successful, were in little villages where the Temple fed us, and froze me. They hadn't the sophistication even to spread straw as insulation on the stone-cold floor.

Strangely, most had electricty; but I found the interminable delays over getting actual volts to the music rather tiresome. I shouldn't have thought that it was a profitable trip, either 'collecting-bag' or 'spiritually'. In a way I was sorry because it was all an intense affair for the Guru Baba. He fasted all the time. The second day he was persuaded to relent and have one banana - which sounds ridiculous, but it wasn't really. I was unhappy on his behalf. And I must say that I could have done with the feeling that the villagers were getting rather less music, and a little more teaching. I found it all a bit sad.

But looking up from Manali the pass looked beautiful.

Thinking that it was the end of the broadcasting, the journey seemed worth all the trouble. I offered to put us all up in the cheap little hotel that seemed to mark the end of civilization. It had looked very cold and short on comforts; but warmth, a hot meal and a bed would be just right for all of us, especially for me, not being quite so young.

The guru, however, was evasive. Somehow the music box got taken down a nasty little dirt-and-snow packed alleyway to, of course, another Temple, this time more or less overrun - quite incongrously - by the sprawling dejection of the Tibetan refugees' corrugated iron huts in that ankle-deep mud.

Unfortunately, this Temple also had electricty - at least, wires hung from the cobwebs. The prospects of any creature comforts faded.

I went off in search of a café, and returned to find a pot-session gradually taking over from the music. This I took to be some kind of reaction to, or celebration of, the journey's end. It was the first and only time on that trip: but they all got badly stoned.

A friendly young man who had accompanied us, rolled himself in the blankets he had brought for me; they lit a huge fire in one of the 'backside' rooms from firewood I had bought to last the night. I had bartered for this in the dark from the Tibetans, who had crawled out of their three-foot high huts to see what I wanted; trying to help the stranger who was floundering between their hovels.

My three co-travellers were asleep when I next returned, lying covered in falling ash, around the fire in the space I had swept out wondering where that revolting beggar had slept last. I had had to buy a 5 rupee mattress just as the market was closing - not seeing myself decently able to get the fellow out of my blankets. They had collared this mattress, while I'd wandered off, dumped it near the fire so that it was burnt and charred into a funny shape, and an inert body sprawled down the other side of it. With that went any last fond thoughts of comfort of any kind.

I trudged back to the closing market, through the trampled snow looking for something, maybe like old newspapers, to sleep on or in but not seeming to make my purpose clear to the few puzzled stallholders.

I got very fed up and annoyed, and eventually turned disloyal - sleeping in the little government Rest House blanket-store, together with the friendly old chokidar [caretaker]; all very unexpected, down by the river (twenty pence, no bribe, but a lot more floundering about).

The unpleasantness, which I half-expected when returning in the hurt of the morning, never materialized. Any upset feelings had evaporated, leaving just the dirty floor and the even dirtier standpipe in the Tibetan mud.

I went back to Pandoh on my own, but they accompanied me as far as Kulu - about half way - and well away from the cold, snowy Manali.

I still had a five-and-a-half hours journey on the bus, mostly on half-made roads - more dynamited out of a vertical cliff-face than smoothed out. I wondered whether the man who'd done it had peered over the edge to see what he'd done. The river was far below, the old valley road having recently been flooded for the irrigation scheme.

I was terrified. The overcrowded bus kept rocking like a boat, something hitting something underneath, three or four feet from the unmarked edge. It kept having to reverse, much of the tail hanging over the edge, inching backwards, stopped only by signals from the conductor's whistle; and then a puncture.

Then suddenly, just up the road, a little settlement and two very civilized young Indian doctors invited us to an elaborate, clean-looking, real, delicately-prepared Indian meal. Illustrating how this whole area seemed full of contradictions; slums and comfort, electricty and cow-dung... incongruities peculiar to this strange Himalayan Kulu valley.

The Two Himalayas

Fortunately the road had just started to leave the cliff-face; otherwise the meal would have been rather wasted on me, to put it mildly! and although the obscurring snow did not seem so far behind, a descending cloud of dust half-obscured the 'road' that zig-zagged in front of us as we descended bumpily nearer the more sensible lower slopes. This reflected the headlights back at us, obliterating the edge. I cursed that following breeze.

I gave a fervent pound note to the driver when we got near Pandoh. Probably three days' wages. I wanted him to know my gratitude. I noted that all the other passengers seemed to be weirdly fatalistic. Something to do, I believe, with their religion.

I was glad to get back to Pandoh. I had been welcomed by half a dozen sadhus, whose low voices guided me up through the jumbled boulders in the dark. But before I had settled down to that nicely-washed feeling the next day, four of them had vanished in their mysterious way. So that I was half-alone in the Temple with a log fire and a sadhu cooking (lovingly, I felt) out in the porch with his 'friend' - if that is the word that describes the vital-casual relationships of these wandering lives.

I was in the mood to look at more of Ulli's long-unposted letter. It would seem to vary from something of an inspiration to just nonsense, according to our natures. But I am pretty certain that if you were fascinated by a near-mahatma, and fascinated by the setting, then you would feel it a sacrilege to quibble; and possibly, like myself, feel it an arrogance to condemn the awareness of the East, even if we do not understand. In fact some of those rather strange pages jolt one almost East of East - and I'd better keep off the subject; except that I couldn't escape from his 'Dad, can't you see?' appeals for the re-direction of my life, Indiawards.

I don't think I noticed, at the time, that one of those unposted pages referred to 'Dad dozing beside me: new dog biting my foot.'

Reading that again now, I once again wonder at my half-learning about foreign lands, and my conscience stirs uneasily. But I slept on, while he wrote those words; and the dog sickened . . .

The problem of getting home loomed up. A 'luxury bus' was talked into existence - their way of pleasing - a common Eastern foible with the truth. But I began to feel that Ulli could stay in India without coming to a horrible end - perhaps in the Temple, on and off.

Ulli wanted me to stay. I strained my back, so I had to. My British employers got their doctor's certificate, in Hindi.

I didn't finish reading all his letters until I eventually got home. They continued, some of them, in that mystifying, half-comprehensible strain. I cannot in truth get my mind round some of it: but that's not to say about others - younger than I perhaps.

One doesn't exactly have to levitate in the mind, but perhaps hope for some rarefied grasp of the meaning. 'When everything that is *mine* is gone, there's no *me* - but just *every*thing.'

Looking back, the Temple and Imperial College, London, do seem to stare at one another, mutually uncomprehending, like the paradox behind that strange sacrificing conception: how, in emptying your life, you might fill it.

It is a long time ago now since the "Dad, I want to go", and the sleezy pub - that unlikely quayside, where the sails were being ruffled by that wind . . . but that's coming

The Two Himalayas

back to Drummond, who seems to understand them both - where they both live. And who understands the wind.

Eventually we both left the Temple. I remember his, now quieter, smile amongst the crowd as I caught the bus for Delhi and home. He was a little sad, and so was I; although he was now more concerned about taking care of his health I worried about the heat in Delhi - the city seemingly a magnet drawing him from those impoverished northern villages. I also worried about the twists and turns of the guru's mind.

Yet I remember feeling that Ulli was more concerned with something else. He wanted to come to terms with that whole cloud of spoilt hopes and disillusions, with the hurt of his cherished pre-Geldof ideas; with how those few (but very few), half-remembered schoolboy bits from the New Testament seemed to have rescued him with a fascination of their own. I half heard, half sensed them, repeated in that special Temple quiet.

I indeed sensed something like that, watching from the edge of his struggles, knowing that he had been figuratively kneeling, as it were, amongst those muttered thoughts, those vividly Easternized texts in his own Holy Land. Mentally kneeling, while all the euphoria over his East/West reconciling dreams collapsed around him. So that, in a way, a quite sadly ineffective 'Geldof' became a kind of wandering Franciscan sadhu. Which puts obedience where spiritually it belongs - strangely powerful in unscrewing our lives; sending us, as it were, on our spiritual way.

A wandering sadhu? A wandering Franciscan? Take up Thy Cross?

But he didn't want to get ill again, nor to starve again.

Couldn't he perhaps try odd translating jobs to earn his keep? He'd A levels in chemistry and physics. He thought he could look up one or two chemists - any that were to be found. They were educated and intelligent men; surely they wouldn't be all that much rarer than doctors? But that, he didn't know about; except he thought they'd probably come up with something or other for him.

He thought, too, that he'd find a welcome in any Temples that he came across. Maybe sleeping in the room at the back, doing a bit of cooking. Perhaps that way he could gradually become a person who was dedicated to men's real spiritual needs, without forgetting their worldly needs?

And it wasn't wrong, was it, to work his way towards Rishikesh? If only to see that very gracious family who had been so kind when he was ill, before he ever found Pandoh. He wondered whether his father back in England had ever got round to sending them that watch, that - as with so many Indians - they would have treasured so much.

It wasn't wrong, was it, to work his way there? Because he had been such a dreamer lately. Dreams that somehow seemed to hallow all his life, not just a part of it . . .

10

When Rishka Lakshiman suddenly left the Temple at Pandoh, she hadn't in fact been able to face the long journey to the place called Rishikesh. Ulli's hunch hadn't been far out - she'd not really wanted to be too far away.

As if that helped.

After all, she'd have to go back through Pandoh unless she was to stay in the Kulu valley for ever. She could look out of the bus up at the hillside and feel sad, not really wanting to pass it; but she did eventually watch it disappear.

She took one or two bus-rides, trying to work out the direction - at least the direction - towards this Rishikesh. Fortunately her little bit of money had not been stolen, even though she had had to sleep out for one or two nights - nasty nights - longing for someone to look after her, feeling horribly vulnerable. And it was still not stolen.

One or two people offered her a corner for the night in return for a bit of cleaning or sewing - how much she didn't question. Maybe even they were just a little concerned for such a fragile, attractive creature wandering the streets in this unusual female-tramping way.

For quite a long time she stayed with a family that ran a kind of shop or store, where there was lots for her to do. They were a bit different, a bit serious, with a kindly, homely streak.

One day - which she was never to forget - there were voices in the shop. Apparently a foreigner had arrived. This she could work out, even though she was sitting right at the back where the chickens were making a noise. Being a family that kept to old traditions and customs, they sat him down in the front room of the store, and tried to please him with their little ceremonies - sacred and meaningful to them, though puzzling to him.

They sprinkled scent on the jacket that he wore loosely over Indian clothes, then gently took his shoes off, washed their hands, made a cup of some special tea in a special cup; and they took two little dabs of cotton-wool which they picked up on the ends of two little wooden sticks; whether these bright, shiny sticks were kept specially for this purpose I do not know but, rather incongrously, even a little bit embarrasingly, they were used to gently push the cotton-wool tufts on to each of the guest's ears - the top back fold of each ear, very precarious, but all somehow very impressive.

By this time, poor Rishka, quietly in the next room - indeed knowing her place in the next room - had realized who the young foreigner was, and was overcome with confusion; knowing only that she was confused, that her heart was beating beyond any sense or rhyme or reason; and that the tiniest glimpse through the curtain was all that she could dare.

The Two Himalayas

Of course her heart subsided; but only in one sense. Confusion spread through all her thoughts, all that little bit of peaceful, half-settling down of these last few weeks. And, of course, when he had left, she asked all she dared about where he had come from, where was he going to, and such-like.

The whole thing left her more and more convinced that he could surely only have come this way, be going this way along this road, if somewhere in his mind was the idea of seeing her again. It made her ask herself over and over again whether modesty - that special modesty and lowly uncertainty of every aspect of her whole life, her whole wandering - whether too much of it would be even fair to him. She recoiled from pushing herself, and yet would have liked to have fallen over accidentally in his path. From her little glimpse of him, it seemed he was a bit thinner. Secretly she would like to have looked after him.

But she couldn't. She swung from there to alarm at the thought of having guessed it all wrongly, a fear that it was all in her imagination. From somewhere in amongst these thoughts, the idea emerged of actually catching a bus to the next market-town, hoping of course that he would somehow have got hung up there doing something or other; trying to earn a few rupees - or even (could it really be?) looking around for her.

Oh God! But she panicked, and caught the daily bus to the next reasonably sized village where there were a few shops and the chance of a job.

She remembered that Ulli had told her - if she had understood his patient, slow English-cum-Hindustani properly - that there were two ways for him to travel. One as a foreigner, perhaps calling on the local schoolmaster, doctor, or chemist - finding a job here and there and so paying for his bus-fares; the other as a sadhu, towards which life he had lately been more and more attracted for reasons he couldn't quite explain, but where one was not asked for fares - the traditional Indian sympathy for holy men, at least away from the bigger towns.

Normally, she could have left it to one of those abiding mysteries of Indian village life to tell her whether Ulli had come this way as a foreigner. As a sadhu - well, that would be a little different. She was very much in the background of this village life, and altogether much too shy a young girl to feel she'd have known of his passing. Nobody talked to her anyway, here in this strange new village.

Possibly the young Sahib (he was quite young for a Sahib wasn't he?) would catch a bus too? Possibly something magic would come from that, quite puzzling, funny ceremony she had watched through the inch or two gap in the curtains.

She set her very superstitious mind on that and felt happier, because the people at an address she'd been given took her in that night, and she hoped that she could soon get a job somewhere else.

Not without a new kind of worry, that little job 'somewhere else'. A fairly well-to-do family in the village seemed to be thinking about it, but for some reason they had asked if they could keep the 'reference' that her father Lakshiman had passed on to her. The ostensible reason she was given was that it needed to be translated. It had one or two longish words. The real reason was that one of the menfolk had taken a fancy to this attractive young wandering girl. How clever he was! Hadn't he been stationed with his Army unit near that poverty-stricken village where it had been written? Stationed there to help the local police? Because, officially at least, weren't the 'criminal tribes' around there? The clever bit was this: how was anybody to know that it hadn't in fact been stolen? At any rater, one always had suspicions about those villagers ... A pretty young girl might well have trouble in ever getting it back again, that is unless she ... ?

And there some instinct, something about his over-friendly manner, started her worrying. Had she perhaps got herself, however innocently, into a not very pleasant trap?

That all fitted in with her daydreams - that *He* would suddenly appear, and would sort out this man who'd kept the letter. She had heard him once, having a go, back at the Temple when somebody had swindled the guru. Perhaps, she thought, he'd best arrive as a young English wanderer after all? Although somehow she preferred to think of him as a sadhu, this idea of a young Sahib rescuing her kept running in romantic inventions through her mind, and carried her far away, almost into a sort of fairyland.

But it was to the mysteries and ways of village life that the story really turns. Because Ulli had, in fact, tried his idea of seeing a village chemist for the second time since leaving the Temple at Pandoh. He remembered having been quite astonished at the sophistication of that chemist's shop up on the hillside above the otherwise dirty-looking shops in Pandoh's high street when he was there. A kind of oasis of modern stock, and everything clean. Quite obviously this would be a good starting-point for his idea of trying translation to earn a few rupees.

It was more of a coincidence that this chemist, well-known for his educated brain, had only a day or two before been given the English 'reference' letter by the over-friendly character who was to call back for it in a day or two. Asking for a translation of one or two longish words like 'conscientious', when in reality he only wanted the delay because it would make the young Rishka realize that he'd got her in a corner.

Not unnaturally, the chemist, finding himself talking to this friendly young Englishman, fished out the letter from behind the counter and asked him what one or two of those long words meant.

For quite a while, Ulli rather puzzled the chemist. Staring out of the window, then back at the letter, then hesitatingly all round the room - yet not at the room. Not just looking for something to buy, but apparently plunged into puzzling, distant thought.

It didn't make sense - and yet it *did* make sense. Long words, yes, but that handwriting? He'd have known that, even without the signature, and the 'Lt. RNVR' - his Dad, years back in the past.

Coincidences can of course be ridiculous; but this was more than that. Or less than that? The Temple world tended anyway to lend strange interpretations to this life ... which all seemed slowly to overtake his mind.

The name 'Lakshiman' to whom it was addressed carried him into a quite different line of thought. Wasn't the real connection not that of luck, of rather shallow gawking at coincidences? Wasn't it more that he himself, this girl, and - years before - his dad, had all flown, but quietly, in the opposite direction to the way people usually went, or usually thought? Wasn't *that* the real, the underlying thread of all this? Wasn't it perhaps *that* which might unravel a story that the story-teller would just bridle at and change around a bit? Whereas in fact there was, as it were, shining through it a kind of meeting-point - a sort of junction of the deeper things of life - from which it could only really be resolved, really be sacredly believed in?

The Two Himalayas

Mixed up with all this, inextricably, it seems were other things, perhaps dreamier things. The feeling that this attraction for that strange girl just would not pass. It had, he felt, got him to where he was; and, he very much suspected that something of the sort - perhaps some horridly sad thing - had a lot to do with where she had got to as well. Wherever that was.

All of a sudden he realized that the poor chemist wasn't exactly clear what it was all about. And so Ulli apologized - indeed rather confusedly, but nicely. And the chemist told him what he knew - feeling that it could hardly be all that of a secret, confidential thing. And so it was that Ulli saw that he didn't need to go much further; and knew that, however it had got to the Chemist, and for whatever motive, the reference was going to be given back to Rishka.

Whoever this somebody was, who presumably was thinking of giving her a job, he came across as not the most trustworthy type of individual. Ulli wondered whether she could be simple enough still to be unconcerned about dubious men like that.

And now that he felt she was somewhere in this village he forgot about buses and jobs and sat about for a day or two, amongst cups of tea and the other deeper things about life that began to invade his mind. On the one hand it all appeared as a sort of spiritually-logical Franciscan QED - a vague, but powerful fate working itself out, or perhaps even waiting to be rejected. Wonder-making, in any case, and making him wonder whether he should go back to the Temple - for her sake. Or whether he *could* just go back, and if so, for whose sake?

Very muddling. Three people had, somewhere along the line, done a 'seek ye first'. A 'seek ye first the kingdom of God'; and in fact 'all these things had been added unto them'. That, at least, was what it felt like. So what the hell? Didn't he believe something like that, anyway?

And as I write this, I cannot quite fathom out whether in all this there was not strange interlocking proof of those convention-wrenching Drummond ideas. Ideas that surely had first thrown convention aside over that young Lakshiman. Which is in a way what I mean by a 'sort of Franciscan QED'. God having his say?

What it means is beyond me. All the echoes as between our two lives - Ulli's and mine - seem to come into it. As if our paths were not miles apart, but closer, mapped one on to the other. God only knows.

And then there was Rishka. Poor girl; yet, if his instincts were right, her own dreamer. Wasn't she, her vulnerability, a bit daft? Getting these impulses towards some restless, vague travel-dream, and no doubt a restless, vague escape-dream. What the hell! Made him feel caught up somehow in a kind of responsibility: a responsiblity for either making her see sense and swallow it all, to get it out of her system - or back into her system, rather, until she choked down the bitterness of her fate and gave in; and felt her two parents watching her choke it all down - while she fetched and carried, and wore out those kitchen pots and pans with continual stirring and some God-knows-what sort of husband . . .

And the thought, thoughts like that, made his lips tighten till he started wondering how he ever could leave her alone; not to be part, somehow even feel part, of this thousand times wonderful - thousand times unusual - challenge that seemed to call, as it were, on every fibre of what he knew was his true self perhaps and very probably his true destiny. To make allowable those things the world deemed unallowable.

His thoughts ran restlessly this way, until he found himself wondering how much of his feeling was more a kind of chivalry than real love. Which thought disconcerted him, and made him suddenly furious with himself, so that he got up and decided to find out where she was - in which shop, or whatever nasty place she might be at this moment working endless hours - endlessly, perhaps, sewing in some back-street, dark little abortion of a workshop. Which all got him searching, until in fact he found her; when her shyness and almost tearful confusion told him to be very careful. So he didn't perform the big knight-in-armour rescuing drama that was very near his heart, but looked at her, very kindly and a bit sadly while she went on sewing, and sewing; and, he wondered, to return without him, one day, to back-street gobaring again?

Not very true to some of his instincts, he fawned rather ingratiatingly on this tailor-cum-shopkeeper whose place she was in, so that he was offered a cup of tea. While he and the shopkeeper sat together, Rishka stayed at her work in the backroom. He made up some story to account for his having seen her somewhere - in the Temple, perhaps, at that Pandoh place? And the man seemed as if he had got the right reply from Rishka, speaking through the gap in the curtain.

Wangling the talk round to whether he might perhaps earn a few rupees by doing, say, occasional English translation, the shopkeeper hadn't been very long in getting round to the rather overdue 'reference', which this possible 'employer' had not yet returned. The shopkeeper hadn't been sure that he could keep Rishka for very long in his shop, so he'd suggested she take the reference to him and ask for a job.

Ulli felt that this was all getting a bit tricky, so he side-stepped all the questions, saying that he'd be around for a few days anyway, and with a wink at Rishka, would maybe try and sort it out somehow.

This little wink meant everything to the poor girl, so that her now-smiling face through the gap in the curtains prompted the shopkeeper to invite her into the room. Ulli felt it couldn't do any harm to say that he thought she was very nice, and that he wouldn't want her to be left with a problem. All, of course, in a mixture of Hindustani and English, and the gestures that get round the difficulties of grammar.

In order to see that she wouldn't be left with a problem, he quietly went to see the chemist the next day and explained something of the situation. He asked, as casually as he could, what sort of a fellow this man was who'd asked for it to be translated. To which the chemist replied, rather indirectly, by asking whether the young girl was attractive or not? Because it was a funny world - or some such remark - and he didn't in fact like any of that too-rich family.

Ulli thought it best to say that he would call back to see the chemist in a day or two, and that if the so-called potential 'employer' hadn't come to fetch it by then everybody would agree that Ulli could somehow find an excuse for getting himself involved - perhaps taking it back to the man. A bit complicated; but really what they both wanted, Ulli and Rishka - a chance to be seen together collecting the thing.

Her knight on horseback (or, I suppose, its Indian equivalent, if there is one).

It all sorted itself out on the young man's doorstep, where Rishka was predictably asked if she'd like to come to tea; hoping for a bit of co-operation from the Englishman

The Two Himalayas

– who ended up by slamming the door. Which, of course, threw him and the girl just that much closer together, in an indirect way.

And it wasn't at all easy for him to decide whether to tell her about that shatteringly emotive complication - that it was in fact his own father who had written the letter all those years ago. However tempting it was - and it was - he kept back from the precipice, because it could work out so unfair, and maybe so unkind to her.

There were copious enough layers of other questions that he felt ought to be faced first before setting her head spinning with that one. Time and again he had worked through how, after all these years, such a few years' age difference between them could have been possible. But, time and again, the thing came out right; and left him with a tempting, provocative, puzzling mixture of the sexual attraction he felt, and of the vast difference between their civilizations.

He supposed that she could learn; would one day even be able to hold her head up if they - as he dreamily wondered - were to visit or even live in England. For instance, she had (he thought) never had a shower or a bath. He, for his part, had no idea whatsoever how ignorant, complicated, or maybe just how submissive such a village girl would be. Intriguing, but a bit worrying - especially with the language problem; with not knowing what mothers told daughters, and so on. I think, though, that on the whole it intrigued him; made him feel that there was scope for love and kindness; gentleness and humour being things that he didn't think he was particularly short of. One of the troubles was that it could well be an utter cruelty if he started off too precipitately and then found it wasn't working, that he was getting unhappy.

Many courses would be open to him, while she probably had no alternative but to return to the mud huts of the village or to the more dubious parts of the towns. A young girl might easily wind up with no alternative - the thought of which horrified him, and frightened him badly.

So what was he to do? How was he to start? Where to go? Where to stay? Where to find the money for two rooms? Or would he have to spend his nights out on the pavement? And all the time not unaware of that lovely young body. Hell! wasn't she lovely, really lovely. That would be ridiculous; except wherever could they find some kind of answer?

He did actually come round to the idea that she could have the bed, and that he could curl up on the floor somewhere, waiting perhaps for it to get dark before they tried to settle down. Whether either of them would or could sleep would be another matter. Not quite what is natural, but perhaps he could suggest it to her, somehow or other?

After a few more days, mostly at the shop, of keeping up their *status quo* of vague friendship, he decided that enough was enough. He took her over to the little coffee-shop and told her - tried to tell her - what was in his mind; because he saw just no other way. They would set off, and he would sleep on the floor.

I don't in fact know whether Indian girls blush (and perhaps I should beg their general pardon) but I expect she did. I am rather more certain that she let him hold her hand, and gazed at the sincerity in his face, and just hung on, while her poor mind bent beneath the rush of peace mixed with storm - that in fact was trust mixed with self-doubt about her village self.

And that is what they did, a day or so later, in another village where they managed to get a room. Not that civilized know-how came very much into it. But there was a bucket of cold, or rather tepid, water of uncertian origin he thought; but he was pretty used to that sort of thing by now.

And he had a little wash. Not too extensive. But after that he half-pretended to have a little doze in their one rickety chair. Except that he was wondering whether he really should have used that piece of soap, and whether she might too. Perhaps she might. He was quite sure that she would be quite trustful about his "Me no look".

A sudden wave of confidence made him stretch upon the bed; perhaps it was from watching her comb her lovely hair by the window. His idea of sleeping on the floor suddenly came to him as too awkward; more than a little likely to keep up a kind of strain, and in any case, hardly the best way to engender real trust. Not *real* trust. That came best, didn't it, from closeness that was just close?

And so they were quiet and were close. And it was a long time before they found themselves looking at one another. He could stroke her forehead, couldn't he? And he did. And she smiled and seemed to be at peace.

She was staring for a while at the ceiling. It was her first time in anything like such a situation, anywhere. Of course it was. She knew, instinctively, that she could trust him. Perhaps tomorrow his hand might move a very little; but it would only be a very little, and she would trust; and find trust exciting.

They talked of the idea of going to Rishikesh. The 'uncle' had been very remote - not a real uncle - perhaps more a 'pretend uncle' to give her journey some respectability, it all being very eyebrow-raising, this journey alone by a young girl.

Rishikesh was one of those places that wandering people landed up in. Her father had wanted to go there when he was younger. Maybe in a funny sort of way he had liked the name Rishka because it sounded of far-off things - things far from the poverty and confinement of his little village in the country behind Bombay.

There was money to think about. The sadhu life, drifting from Temple to Temple - spiritually involved, but doing odd jobs like cooking for your keep - was an uncertain one for a young man with a girl. It might work; it might not. Perhaps at some Temples and not in others. Females were normally considered second-class - for the porch only.

Generally speaking, money was not too much of a problem for a sadhu, but he wasn't too sure whether it would work out for the two of them, especially as he could not go back to sadhu Eastern dress - or could he? He didn't know.

He began to think, however, that it would be more helpful to Rishka if they got involved with town life instead of going off to Rishikesh. It would help her language-learning and also give her more confidence and knowledge of the world, and make her feel less sensitive and insecure about their only partly-compatible clothes; while his hair, which had been shorn, grew towards a more matching hippy-length again.

So they worked and odd-jobbed their way down to Delhi, not always getting stuck without a roof, scraping together a little fare-money, which was partly made possible by the very unusual and intriguing mixture of a not very 'white' young man and a shy, but intelligent, young native girl who was always in trouble with her English.

They had it in mind to go to that very cosmopolitan little hotel in Delhi, where those two officers had taken her, and where the owners had been so kind. This worked out very happily, with pleasant smiles and kindness and an offer of domestic jobs for Rishka, as she had done before; but they didn't quite know what for Ulli.

Probably because of his idea of offering to sleep on the floor, for the young girl's sake, they took to them enough to let them have an odd unused room at the top; not a very wonderful, lettable room, but it was very kind of them.

Funnily enough, Ulli found himself hoping that there wouldn't be a fire in this rackety old wooden building - a hangover from his Western background - while the owner-couple hoped that one of the other older guests wouldn't find it just a bit too 'unrespectable'.

He had little idea about so many things concerning Indian girls. He didn't want to find out about those sari things, and the multiplicity or non-existence of whatever else there was. This trust thing seemed to have found a different part of his nature. That trusting look in Rishka's eyes kept amazing him - kept him so that he was amazed at himself, and how tenderly he responded.

He thought it would be better if he lay quietly, maybe facing the other side, so that she might soon lie near him in the dark, all nicely washed. Perhaps she would turn towards him, shutting her eyes, and not seeming to mind - as maybe Indian girls were not made to mind?

Perhaps, except for the washing, that was what they did in huts. Which would be lovely, except that he resented any idea that a woman's feelings counted for nothing. And here they did; around here at least.

In truth they were both wondering about this sari and the night-time. All this speical bedroom business, the electric light, and this sari - how it sorted itself out. It was a plunge into a strange world to her. But she seemed to hesitate onward from one little bit of trust to another.

The wild thought even came to her that very soon - perhaps tomorrow even - she would have a wash in the secret corner, but suddenly come out and stand before him; like you didn't do in huts. That scared her a bit; not knowing what white girls did. She wasn't at all ready for thoughts like that. But was he waiting? Would he think it just lovely? Or would it make his English heart all confused?

She wouldn't have know what to do with an Indian man, but presumably *they* would know all about that. And perhaps Englishmen - the nicer Englishmen - only like shy girls? Perhaps that would wreck the whole thing, if she did that, and it went wrong.

This washing was very nice; made you feel nice, especially your hair. But next day you felt dirty, which you never used to. And so you had to wash again. Perhaps soap made your skin funny? She suspected unwashed European girls smelt like Indian village girls didn't.

Her hair was very beautiful, very shiny now, very long; well below her waist. She suspected - and secretly hoped - that one day very soon he would go on stroking it even though it might have fallen a little immodestly. She would have liked her sari to have been sometimes more carelessly arranged to that his hand could fall like her hair.

Then, suddenly, she would react from such thoughts until she felt like running way - not from him, but from that sea of doubts that came and went like the tide.

Ulli kept on with the idea that some of the odd pamphlets, brochures and other such publications could well do with a few corrections to their obviously badly-translated

English. He managed to get an easy part-time job with a local printer, so that for some weeks the two of them stayed on in the little hotel.

He spent a great deal of time writing to us at home. Really working out his own thoughts: about sadhu life, about Franciscan ideas, and about his interpretation of the Hindu philosophy of the temples, and so on.

Although for a time he had abandoned the complete sadhu way of Eastern clothing this had really been in order to get around more quickly during his search for Rishka, and he now tended to revert to it, because it fitted in more with the thoughts that he was trying to formulate; with those ideas of the not-so-materialistic interpretation of things that had drawn him, and so many others in those years, out to the East - basically restless, because the West seemed to have lost something powerfully relevant to man's real happiness, which the East hasn't quite lost; and the idea of the West needing the East and the East needing the West, which he was always writing about, was very largely the basis for his 'tourist bungalow' idea anyway.

This idea, to which I had also been drawn, was very much in both their minds. In perhaps different ways for each of them; and I think it was never very far from their hopes for the future they looked to, a future which they knew would be difficult, but where many of the difficulties could be solved by just such an independent project - independent of either rigid Eastern or rigid Western ways.

Perhaps it was, even for their young generation, in advance of its time. Sadly, he was never to know how the young of more recent years, running for food for the Sudan and Ethiopia, were to find themselves strangely half-elated, born on a wave from 'Live Aid', surprised by an hour of fulfilment by 'Don't they know it's Christmas?'

During this time, he returned for a few days to the Temple at Mehrauli, near Delhi, and left with the impression that here at least was a community that would not frown upon this Rishka affair. Fine and wonderful people, who were regarded (one or two of them) as 'mahatmas' - the highest of gurus. In fact, they gave him a sense of security about his problems. He told me that they had so reassured him that when he left, a song from *South Pacific* had, incongrously, kept reverberating through his head:

Before you are six, or seven, or eight they'll have taught you to *properly* hate.

In the show the US officer discovered he couldn't marry the native girl. Another man's dream of a girl to share his island was shattered. They may have taught others 'to properly hate' - which filled Ulli with the idea that he could show them - all of them - how to properly love.

But when he arrived back at the little hotel, Rishka was nowhere to be found. The kind lady especially was very concerned because she liked the girl, and was an unusually sensitive person. Ulli spent part of the night sitting on the bed in perplexity, worried as to whether she had fallen victim to some mugging or similar situation. Then he went around the streets for hours. Over and over in his mind was the churning thought that he might have done something wrong, or been insensitive. He felt sick; and he wondered time and time again whether perhaps he had got himself into one of those situations which a mere male cannot understand. So real, yet so - to him - illogical.

When he woke up from an exhausted doze, he moved the pillow, and there lay the answer. He thought perhaps he would find that she had written a few very simple words. A few simple words? All it said, a sad little scrap of paper, was 'Me gobar girl'.

The Two Himalayas

It couldn't be anything to do with doubts, deep down, that after all 'they may have taught him to properly hate'. No. Not that Sahib. That was what the world thought. They trusted one another. Rishka knew that the world could hate. The world could do what it wanted, but they were somehow on another planet, and they both knew it.

But it could have been something remotely deriving from pressures, real or imaginary, on both of them. Him for her sake, and her for his sake? Possibly, more that she was dreading pressures in the future upon him; pressures that he was gladly defying now, but that she saw might one day surround him with a thousand restrictions, to be borne one day just for her - for his one mistake of caring for her.

And so on and on, and round and round. Contradictions. White girls; dark girls. Her sake; his sake.

Except he felt sure that they really loved one another. He knew, and in the end felt it quite finally, definitely, that she had left, not for her own sake, but because she felt that she might one day be a burden to him.

And then, perhaps not so surprisingly, came another refrain from that same film which had moved so many people back home, and that he had even seen twice over:

> Some enchanted evening,
> When you find your true love,
> When you feel her call you, across a crowded room,
> Then fly to her side, and make her your own,
> Or all through your life, you may dream all alone.
> Once you have found her, never let her go;
> Once you have found her, never, let her go.'

Not too specially applicable, perhaps; but . . . 'never let her go'.

And so he got up, with all the streets of Delhi as his hunting-ground.

Assuming he did find her, wasn't it time that he told her that the letter she carried was from his own father? Perhaps that would seal the thing as almost the working of fate, of some God rather?

This made him, not for the first time, work out that his father's vaguely-remembered story of jumping off a train on the way to a hill-station, at the end of the Second World War, must have concerned a place near Bombay, a little station on the line from Bombay to the hills - the Ghats, no doubt. It *had* to be. Maybe that was the part of the country she had made for: her village home, to try and sort herself out.

He decided that he would first wander through these Delhi streets, combing their narrow alleyways as best he could. And if that didn't work, he'd try and find that village. After all, if she *had* gone there, she would no doubt have stayed on for a while. So, Delhi first.

He had a vague memory of her story, of how she had landed up here and had been advised to sleep out - if she had to - in the big lit-up square, which he would try to identify. Presumably fairly near the main railway station, he thought.

The square in which he found himself had lots of nasty little alleyways leading off it. That night he wandered around them, hoping to God that she wasn't somewhere in one of them. They struck him as being, to say the least, not quite the place for an attractive young female, village girl or no village girl. And he got really worried. The next day he thought he'd have a look in the cafes and little shops, the stores and whatever. Never mind what the natives thought! He was going to look, and look properly. He wore something like European clothes with his Western jacket, so better to attract attention, maybe even indirectly, *her* attention, through the idle chatter that would no doubt follow him round.

In one of the cafés, fairly early that next morning, he saw that they were cleaning the floor. God knows! he thought, how often they did that. He was a little bit put off to notice that blood was coming from the foot of the girl who was kneeling, with her bucket, cleaning the slabs of the floor. Not very much blood, but enough to make quite a bit of her foot all red; not quite the sort of way of going on that one could approve of, but one has to suppose that this is the sort of thing one is going to see if one pokes one's nose into these places. In a way, he wanted to go inside and belt the owner, probably some nasty, greasy little type who could do with a thick ear for his pains anyway.

He didn't get a thick ear, but just a few words. Which turned Rishka round, so that their eyes met and he looked at her foot and at the nasty little owner-type, and his feelings welled up - so that the bucketful of dirty water was thrown over all the food that was spread out amongst the flies on the counter - not exactly his old 'peace and love' theme, more like our old friend prejudice, except that chivalry kind of busted the situation!

She got up. And the story told itself. She had hardly been able to walk, and, hungry, had begged this 'almost Gobar' job. Nothing could have infuriated him more. He lifted her up bodily, and to the amazement of those around here was an Englishman carrying a young Indian girl, and shouting for a gharri, a scene unheard of in this town of unheard-of things.

Heaven knows what she thought! But heaven did know what she felt as she let herself be lifted into the cab, wondering at his strength - at which he was wondering too.

Of course they drove back to the hotel she had left. And another youngish girl who worked there came shyly up to him and said, "I write letter - Rishka no can. She very sad. She love you. I think she go home. I tell you, if you want, where is her home. But I hope she stay with you. She love you. Indian people like you. It come out alright", and disappeared into the kitchen.

They both settled down after all that. He bathed her little foot and kissed it better - which all could have been the best for both of them. Having been carried around pleased her, too, in something like the same way.

The Mehrauli thing helped, giving her the feeling that there weren't quite so many ill-disposed folk in this world; that they might find others around who wouldn't be unkind or make difficulties for him. That was very important, very near the bone as far as her running away had any bone, because it had all been due to plain, mounting, bewildering doubts.

And it was very near the bone that somehow that carrying seemed to be melting away the last physical doubts.

Although they stayed on for a while in the hotel, and they found themselves growing towards one another, I suppose their youth made them want another setting, where they could grow together but also achieve something of a joint venture.

The Two Himalayas

He was intrigued by the far-away look in her supposedly village eyes when they spoke of that 'tourist bungalow' idea: how it wouldn't seem to go away; it seemed to be positive, rather than negative. A goal rather than a state of drift. So they eventually set off together for the old Temple at Pandoh, where the idea had been born, and where there was that little triangle of stream-isolated land that had fascinated his father at the time of Ulli's illness. So much so that his father had even seen the District Commissioner about it.

Pandoh was, of course, where they had first met the week that had ended in her running away the first time; towards which she had gazed on her rather in-love-bedevilled journey back from Kulu, and where, unknown to her, Ulli had been ill so that the Temple folk had called for his father to come from England.

Wasn't this the place - just the place - for him to tell her, at long last, about the letter? He had so persistently, perhaps so unkindly even, kept the letter out of it; leaving it to their natural mutual attraction to knit them together. This mind-boggling thing about *his* father having written the letter that *her* father had treasured, and passed on to her, was too romantic, too dizzy-making; not really fair. And no doubt both of them could have ended with the nagging uncertainty as to how much of their attraction was real and how much was some sort of mental trap that wouldn't stand the test of time. The worldly cynic can say that judgement is not based on coincidences; if in fact one could be so spiritually cynical as to disregard the attraction of a certain kind of people for similar lines of action - people who tend to jar the world in the same way, so maybe drawing their lives together. Coincidences, but not quite coincidences, perhaps?

Better present this development in his life to the guru, he thought - uneasy in his mind as to the reaction he would get. Although he and Guru Baba Dass had been for a long time much, much closer than father and son, priest and pupil, and Ulli had even been adopted as a 'Dass', he wasn't quite sure how it would turn out about the girl.

Guru Baba Dass, of Ulli's past long spiritual journey - of his letters home, and his night-long vigils - was not quite the most predictable of individuals. But when he - for reasons best known to himself - defied convention and welcomed the girl to his Temple porch, it was almost as if he sensed that something differentiated her from the normal Temple cleaner. Perhaps he had sensed it during the time she had stayed a week, when Ulli had been drawn to her, and had - hadn't he? - gone off to look for her, and had returned ill. Perhaps it was because he trusted that Ulli had some kind of instinct about people. Maybe even because he also secretly rather fancied her himself just as Ulli did and would, secretly, have quite liked to have gone off to look for her himself.

So that, when evening came, Ulli knew that he could take her to the 'backside room'. Perhaps not for the night, but for an hour or two to be near that kind of soft quietness of the Temple evenings.

Instinctively, she let the religious aura invade her mind. And she stared into this religious space thinking about the prevalent belief in reincarnation that was threaded through the Indians' outlook; whether she, as the very lowest caste, could hope for reincarnation after death into another life in a higher caste, she was not sure. But this unheard-of thing that was happening to her in this life - sailing, as it were, in and out of the caste system - was not a little disturbing. Lying here with this young Sahib could not be real somehow. And the guru's attitude - however could that be? Perhaps it was all too much that Ulli had chosen this evening, that was so special, to start telling her how his own father long ago had sat in the doorway of the ramshackle little hut that was their family home and written that reference for *her* father.

The gobar floor, where he had first seen her kneeling at her lowly job, that was Ulli's spiritual home. And it suited him, in a spiritually explosive way, that she was still a gobar girl, and he was once more a kind of sadhu.

He pretended to clean her little hand, and didn't pretend to kiss it. She wept a little, her heart too full, trying to cope with a flooding mind; thankful to she knew not what. Instinctively, still a bit shyly, she gave him her left bosom, thinking her heart and his feelings were somewhere there, which perhaps they still were - although she was and had been all his since he had bathed her foot. And every full moon, which it was tonight.

Why it had all happened as it had - like many things in India - wasn't entirely clear. She had grown up not having to make sense of everything; blaming those village gods that so often accounted for village misfortunes; accepting them, leaving it at that. Puzzled that, for once, those gods were fairy-like in their capturing of all these impossibilities, and bringing them into her life. Unless, she wondered, could it be something to do with that big foreign God? She didn't know; she wondered.

I was never quite clear how long it lasted, this honeymoon between them and the gods.

Not very long, though.

Other gods came out of those mountains that were not all that far away on the horizon. Or perhaps that's what she thought. Because she got ill, and no-one knew why. Then a little better, then ill again.

But this time Ulli told her it wasn't gods, it was the White Man's powder - that milk-powder that she didn't know had to be mixed with boiled water and not the water from the local stream as he had seen happen - Pandoh being one of the world's villages where this lesson had been painfully learnt, so that even those big international companies might have been ashamed.

And she got ill; so ill that she began yearning for her home - for Ulli to take her there - which he understood. An idea that, if it were not for this illness, would have quite intrigued him; but as there was a long way to go could have been perhaps beyond her. The thing that clinched it - for both of them - was that her father had given her a return railway ticket from her village home station to Delhi. Ulli thought he would get over the money question by being dressed as a sadhu. He wan't sure about trains, whether sadhus were never charged, as on the buses.

As Ulli wrote home once to us 'In England we would be bottom class, here we are top class.'

Anyway, they started out - and they got there; at least to the little village station, where, after hours and hours of exhausted misery on the bus, and then a more comfortable, but wearisomely long affair on the train, she finally ran out of strength.

And once again, India saw a young Englishman carrying an Indian girl, this time up the hill (with quite a few rests), but up the hill to her little village of huts and tents. And one or two old folk remembered how another Englishman, long ago, had carried the young Lakshiman, who, being hungry, wanted three annas to carry his case . . .

Nobody was ever quite sure how those next days passed, with everybody's memories somehow sacred for everybody else, watching over her end.

11

And when he had taken his final leave of the villagers, and her family, he carried himself and his sadness to the only place where he might be able to find an echo of his real thoughts. Both from the men around, and from their sense of the God who was always around. At least who *had* always been around during most of his time at the Temple, which was the strange kind of persistent spiritual infection which one got from the guru. He went to Mehrauli of the mahatmas.

One or two of the letters that he sent from there were very sad. He seemed to be struggling, as it were, to add another dimension to his life. Perhaps it was because he had tasted the humbling of his 'Oxfam' hopes that he had needed her humility?

I remembered my little gate, all those years before and that little book; and hoped for him that he might even find solace or some spiritual rhyme or reason in her death - in that he had carried her up to her death.

After a while, he left Mehrauli for a little village south of Delhi where the guru's sister lived, which pleased me. But he was afterwards to tell his mother that a conviction - some kind of premonition - came over him in the middle of the night that if he did not make himself get up and make for home, then he would never see us again. So he got up off the mud floor of the hut in the dark and dragged himself toward the bus for Delhi, and they took him as a sadhu.

We here at home had a sudden call from the British Airways local office; a Saturday, banks closed, ambulance warned. Who was it, I wonder, in that Delhi office whose decency overcame his doubts, whose initiative got Ulli to London? Who prevented - quite unknowingly - the horror of his perhaps lying captive to superstition or to local-medicine end on a mud floor, the 'Rage' at its worst.

But, strangely, he got better on the plane they put him on, and we survived a day or two of rather way-out behaviour, and found a pair of shoes for his nine-months unshod feet.

Although he believed in our UK hospitals, and was hoping for a check-up in any case, we found him - he found himself - becoming a bit of a problem. The more so when his sudden irrational fear of water raised the desperately rare question of rabies.

As is apparently typical, the symptoms were confusing. Hence you get involved with psychiatry - and of course his background and appearance gave enough false trails for those who get rather lost outside convention. And as psychoanalysis is hindered if drugs are administered, this poor patient had to stick it out, unrelieved and mentally torn apart, until the increasing choking spasms proved finally that it *was* rabies; until the repeated confused attempts to get to the sink - confused because he was frightened of drowning in his own saliva and yet at the same time frightened of the water at the sink - until eventually these attempts and most real tenseness was mercifully stilled by drugs, and the gathering weakness.

Without drugs, after two or three sleepless days and nights even the more subdued intervals that come and go are (I suppose) so pregnant with the mental horror of this weird, unknown, mind-blowing world that even these intervals of calm can disintegrate. And hence the foreign dread of mad dogs spreading 'The Rage'.

But there was something contradictory in the depths of all this.

Although he did not seem to be quite able to reconcile himself to any fairness in Rishka's death, nor to clear himself of any blame, he seemed somehow to reconcile himself to his own; to death in general; which was a relaxed, almost a smiling sort of subject, and had been thus from the earlier days of the Temple influence.

And in any case that hoped-for Franciscan wandering life out there, living out his version of 'Thy will be done', seemed to push insecurity to its limits.

Perhaps one could even be forgiven for saying that, as death loomed over his bed, we saw a conviction that convinced; a faith in that conviction being pushed to its limits, and radiating not a struggle but a peace . . . which seemed to prove that it was not all in the world of thought.

There were those few words from his last scribbled letter, which I'll repeat, because they are embedded in our memories:

'Too weak to write: will entrust myself to God - the only thing worthwhile in life. I know I'm OK.'

And the strange originality in the little note found under his pillow:

'He who grieves for me does not know Me.'

'Does not know Me'. . . where did that thought, or more than thought, come from?

I believe that when he realized that a London hospital could not save everybody - could not even have saved her - he began to forgive himself for her dying quietly on the floor, on her gobar, and with her Gods, and went in peace.

> '. . . I picked a big bunch of nice flowers yesterday. It doesn't seem three years since my brother died, in fact I'm not sure what it feels like. To me he is still 'there', and always will be, just like to other people God is 'there' when they're alone or in trouble. Anyway he's part of God now, so it's all one and the same. He's just a bit of God I know personally.'
> '. . . I curled my hair the other day . . .'

Afterword

Some ten years later we sat up late watching Philadelphia, where 'Live Aid' ended up. Which is where, we all know, he would have ended up.

I hope you can see why this book was born - as we were crossing the road from the hospital for the last time - and why I know that this India-Ireland story really belongs to the restless.

There might even be some sort of answer in its downwardly-mobile pages.

Not too much about God, but about yourself: obedience to your real self, where maybe lies your *real* peace - which is the odd bug of this book.

... maybe obedience even to that touch of the Geldof in you, which could 'run the World'.